T3-BNJ-330

The Coral Seas

HANS W. FRICKE

The Coral Seas

wonders and mysteries of underwater life

with an introduction by I. Eibl-Eibesfeldt

G. P. Putnam's Sons New York

Thames and Hudson London

Translated from *Korallenmeer* by
Alan G. Readett and J. R. MacCormick
for ELB Languages Group, London

ISBN (UK): 0 500 01104 4

Library of Congress Catalog Card Number: 73-78487
SBN (USA): 399-11178-6

Printed in West Germany

Photographs by

George J. Benjamin, Canada
Jane Burton, England
Claude Carré, France
Ron Church, USA
Ben Cropp, Australia
Walter Deas, Australia
Irenäus Eibl-Eibesfeldt, Germany
Hans Flaskamp, Germany
Hans W. Fricke, Germany
Peter R. Gimbel, USA
Hermann Gruhl, Germany
Hans-Rudolf Haefelfinger, Switzerland
Sebastian Holzberg, Germany
Bruce Hunter, USA
Siegfried Köster, Germany
Peter Kopp, Germany
Dietrich B. E. Magnus, Germany
Jack McKenney, Canada
David Masry, Israel
Horst Moosleitner, Austria
G. R. Mundey, England
Allan Power, Australia
Ruud Rozendaal, Holland
Ludwig Sillner, Germany
Piero Solaini, Italy
Akira Tateishi, Japan
Ron Taylor, Australia
Valerie Taylor, Australia
Herwarth Voigtmann, Germany

Contents

1 *(Opposite)* Coral reefs. An atoll in the Tuamotu Archipelago in the South Seas. Atolls result from the subsidence of an island. A shallow lagoon is formed in the centre of the ring-shaped reef, while on the outer side, the reef plunges steeply into deep water.

2 *(Overleaf)* The Great Barrier Reef off Australia is almost two thousand kilometres (1,240 miles) long, and is thus the largest coral reef in existence. This gigantic undersea landscape was created by billions of tiny coral polyps.

3, 4 For many inhabitants of tropical islands the coral reef is their main source of food, since they can live only on their harvest of fish. Many coral islands, such as San Blas (Plate 4) give the appearance of being a paradise, but their outward beauty betrays nothing of the struggle for existence waged by the inhabitants.

5 A fisherman catching fish larvae in the surf of the China Sea.

4

5

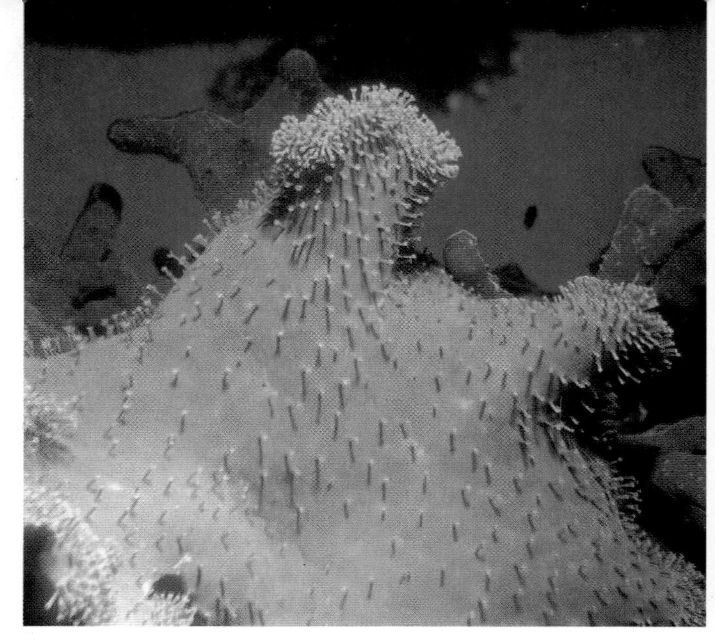

6

7

6 One of the architects of the coral reef is the soft coral *(Sarcophyton)*. The polyps extend their tiny tentacles into the water especially at night.

7 Red algae also play an important part in the formation of the coral reef. They generally occur in relatively deep water or in the shadier crevices of caves. Another group, the lime algae, can secrete lime, which cements the pieces of dead coral branches together to form rigid plates.

8 The staghorn coral *(Acropora palmata)* has been particularly successful in colonizing the coral reefs of the Caribbean. The porous skeleton of this coral facilitates the deposition of lime, and this results in rapid reef growth.

8

9 Some fishes, like the parrotfish *(Scarus)*, can nibble away the hard calcareous skeleton of the coral. They later excrete the lime in the form of very fine dust. Certain kinds of fish thus manufacture sediment and contribute to reef formation.

10 The remains of wrecked ships often form the nuclei for the formation of new reefs. Many creatures otherwise present only at considerable depths are found in the dark interior of the sunken ship, so that wrecks are very rich hunting-grounds for marine biologists to explore.

11 The trumpetfish *(Aulostomus maculatus)* takes a ride on a grouper. The trumpetfish uses the grouper as a stalking-horse to approach small reef fishes, until it can suddenly dart on its prey.

9

10

◁12 A menacing reef-destroyer, the spiny starfish or crown of thorns *(Acanthaster plancii)* has increased very considerably in the Pacific in recent times. One individual will devour roughly a square metre of reef in a month. The scale of the damage caused by the starfish has not yet been precisely established.

△
13 Like any other starfish, the crown of thorns has small tube-feet, which enable it to climb across the coral.

◁14 The white areas in the photograph show where the spiny starfish has been eating the coral away. The pure white base of the coral is left standing, but the polyp – the living constituent of the coral – is destroyed.

15 *(Overleaf)* The discovery of free diving has opened up the reef to thorough investigation. Man can now explore the last untouched life zones on earth.

Introduction: the Tropical Seas

When I was invited in the spring of 1953 by Hans Hass to join his first 'Xarifa' expedition, I had no idea what a new and wonderful world this opportunity would open to me. I had heard a great deal of the ocean, and of storms, sharks and brightly coloured coral fish, but the largest area of water that I had known up to that time was the Neusiedler See in Austria. What is more, up to then I had been almost entirely concerned with research on amphibians and land mammals. And now I was to have an opportunity to dive in the Caribbean Ocean and around the Galapagos Islands. All this is now almost twenty years ago, and my vague pre-conceived ideas gave way to a colourful reality.

My acquaintance with the sea considerably widened my zoological horizon. The multiplicity of coral fishes opened my eyes to the extraordinary phenomenon of adaptation of shape, colour and behaviour exhibited by these creatures. I began to be fascinated by the problems of their evolution and ecology, and since that time hardly a year passes in which I do not visit a coral reef somewhere.

It was around the Los Roques Islands off the coast of Venezuela that I learnt for the first time to know the overwhelming wealth of life of the coral reef. Equipped with a lightweight oxygen apparatus, I swam carefully through these rock gardens, past hedges of coral, growing luxuriantly like bushes, among whose branches lived the dark, yellow-tailed reef perch, *Aethaloperca,* dotted with luminescent light-blue spots. I dived beneath the spreading branches of the staghorn coral and finally came to the steep wall of the reef, a cliff of rounded blocks of coral, which dropped off quite steeply into the depths. Here, some 15 metres (50 feet) below the surface of the water, I sat down. Below my feet, the reef disappeared into the dark blue of the ocean depths. Schools of fish flowed smoothly by me. They appeared out of the surrounding blue-ness, took on precise form and then disappeared again. A shark patrolled up and down ahead of me, stiff and impassive, only his appraising eyes in constant quick motion.

I examined the wall of the reef. It consisted partly of large, overhanging blocks of coral and was fairly heavily fissured. A black angelfish with large golden-yellow-edged scales emerged from a cave, stared briefly at me, growled 'crack crack' and rapidly hid again; another – black-and-yellow striped – took its place. A pair of light delicately-patterned butterfly-fish, each with a striking dark eye-spot below the dorsal fin, nibbled at the coral near me. A green parrotfish bearing yellow spots on the gill covers gnawed at a staghorn coral, as though it had been a turnip. Brightly-coloured slender lipfish and high-backed surgeonfish scurried busily around me with flicks of their pectoral fins. Fish swarmed all about me. The variety of shapes was confusing in the ex-treme, and led me at the time to fear that I would never be able to discern any order or to find my way. And yet my first dive brought me one single observation, which was to determine the entire orientation of my work for the future: while I sat there observing what was going on around me, a large grouper swam slowly by with leisurely movements of his gently moving pectoral fins, a sinister form with a fearsome mouth. Two small and slender blue fish swam up and began to explore the predator's skin. One of them swam towards the gaping throat and – I could hardly believe my eyes – disappeared into the yawning cavern. Well, that's the end of him, I thought – but how wrong I was. The grouper remained absolutely stationary, while the little blue fish swam around in his open mouth and gently removed a morsel here and there. After some time, the client almost completely closed his mouth, and the little scavenger swam out into the open water. The grouper then shook his whole body and the two little fish returned to their coral home. The grouper swam off and disappeared.

Underwater barber's shops

While I was still recovering from my astonishment, up came another fish and repeated the same process. It very soon became clear to me that the little fish were 'cleaning' the big fish of parasites and dirt. There was a constant procession of new clients. Obviously I had by a lucky chance come upon a barber's shop in the reef. And I soon also realized that hosts and cleaners were communicating by means of a series of signals, so that what I was looking at could be described as a piece of organized co-operation. Subsequently I introduced the term 'cleaning symbiosis' to describe this operation. When I came to study the literature, I found that the American scientist William H. Longley had already given a brief description of this operation in 1918, although he did not then recognize the symbiotic character of the relationship.

In the meantime, I have dived in many tropical seas. I came to know the Pacific atolls, the Great Barrier Reef of Australia and many reefs in the Indo-Pacific and Caribbean regions. And everywhere I found cleaner fish. In the Indo-Pacific region these were primarily lipfishes. The blue-striped cleaner wrasse *(Labroides dimidiatus)* was the commonest. This was the fish I observed most frequently in 1958 during the second 'Xarifa' expedition in the Maldive Islands, south of India, where a dive in the Fadiffolu Atoll provided me with a further determinative observation; once again I found myself in an underwater barber's shop. Three sea-bass, one grouper and several impatient butterflyfish were awaiting their turn. Two blue-striped cleaner wrasse were busy about their customers, and then I suddenly observed that a third cleaner had joined the other two. One of the sea bass welcomed him with open mouth. But the third fish had hardly set to work, when the sea bass flinched and suddenly took flight. A little later I caught the presumed cleaner fish. As I held him in my hand, he bit my finger so badly as to draw blood. This made it clear to me that this was no cleaner fish, but a saw-tooth blenny. I had already made the acquaintance of the saw-tooth blenny in the Galapagos Islands. There, several of them attacked a wound on my leg, and I was subsequently able to observe how these little fish made use of their sharp teeth to bite tiny pieces out of the fins and skin of other fishes. But the Galapagos saw-tooth blennies looked rather different. The kind that I had found here in the Maldive Islands resembled the cleaner wrasse in every detail; they had precisely the same colouring and had managed to imitate the behaviour of the cleaners so precisely that I was not immediately able to distinguish them from the true cleaners. They had camouflaged themselves as cleaning fish, and thus deceived the clients. Their camouflage enabled them to approach their victims easily, then suddenly to attack from close quarters, cutting away pieces of the skin and the gills. This particular example showed me very clearly the tremendous competitive pressure which operates on a coral reef of this kind. It is hard to conceive of it being 'economic' for a species to evolve to fill such a 'way-out' niche as that of a cleaner mimic.

I began to investigate other cases of symbiosis among fish. In addition, my interest in the multiplicity of forms of adaptation was aroused, and I began to take a systematic collection of reef fish. I collected more than 420 vertebrate fish species from the reefs along the Maldive Islands, although I dived only to a depth of 50 metres (165 feet); this figure may serve to give an idea of the abundance of species occurring in a very restricted space. I had been given a striking demonstration of evolution at work.

The sea as living space

But how are we to explain this wealth of species along the reef, and how does this densely-populated area appear in detail? To understand this somewhat more clearly, let us begin by talking about figures. Precisely 71 per cent of the Earth's surface consists of water, this entire interconnected system which we call the oceans. The total mass of water is estimated to be 1,400 million cubic kilometres (350 million cubic miles). If all the mountains on the Earth were levelled, so making the Earth a smooth-surfaced planet, there would be a layer of water 2.5 kilometres (1.5 miles) thick covering the whole Earth. Viewed in this way, our Earth is a planet richly supplied with water. The living space afforded by the ocean is many times greater than that of the land, and so much the more in that it is inhabited at all levels, while the land-dwellers are limited to the terrestrial surface.

Every corner of the sea is filled with life. From the eternal night of the abyssal depths researchers have brought up many extremely bizarre species, such as anglerfish with luminescent lures and viperfish which can swallow prey many times larger than themselves. We know of the existence of gigantic squids, with a total length of more than 20 metres (65 feet). More than six thousand different species of fish are known from depths below 6,600 metres (19,700 feet). When Professor Jacques Piccard's bathyscaphe *Trieste* reached the bottom of the Challenger Deep, off the Mariana Islands, on 23 January 1960 at 11,278 metres (37,000 feet), the scientists discovered there a flatfish 30 centimetres (12 inches) long and a shrimp.

Evidence of abundant life has also been found under the Ross pack-ice in Antarctica. In the vicinity of the Koettlitz Glacier, which even in summer lies 28 kilometres (17½ miles) from the open sea, bottom samples have been taken through fissures in the glacier containing representatives of eight different phyla. Every square metre contained hundreds of anemones, amphipods and even fishes. And yet this region is completely cut off from light and the temperatures range from 1.92° to 1.95° C (29° to 35° F).

Whereas, as we have already said, life on the land is restricted to the Earth's surface, in the ocean it extends to all three dimensions. The open water is populated by a multiplicity of organisms which pass their entire life swimming or floating and consequently have developed a whole series of special adaptations which we shall discuss later. The chemical composition of the sea-water is very uniform. In general, 35 grams of salts are contained in 1 kg of sea-water. Ordinary salt (NaCl) predominates with a content of 77.9 per cent and, since the percentage composition of the sea-water does not change, is generally indicated by the percentage of ordinary salt per litre. The salt content itself varies only slightly, if we leave out of account certain side-arms of the sea which have been desalinated by rivers, such as the Baltic. The maximum salt content in the surface water occurs between the twentieth and fortieth degrees of latitude. It is assumed that a major part of the sea salt is leached from the continental land masses by rain-water, from which it is concluded that the oceans were once less salty than now.

The range of temperatures in sea-water runs from –2° C (28.4° F) to +30° C (86° F); in tidal pools the temperature of the water can be even higher. But the fluctuations are far less extreme than on land. The daily and seasonal temperature variations in any region are generally very small indeed. In the open sea they never exceed 10° C (18° F). Around the Equator, the surface temperature is 27.5° C (80° F), with seasonal variations of a maximum of 2° C (3.5° F) and daily variations of only 0.2 to 0.3° C (0.3° to 0.5° F). Only in the vicinity of the coast are the temperature variations somewhat higher. In the middle latitudes, the seasonal temperature varies by 6 to 8° C (11° to 14° F). Below a depth of 200 metres (660 feet), we cannot observe any seasonal variations. The temperature falls off with increasing depth. In the tropics, at a depth of 200 metres (660 feet) the temperature lies around 20° C (68° F), at 1,200 metres (3,930 feet) it lies around 5° C (41° F). In the deep ocean waters, the temperature is only 1 to 2° C (34° to 36° F).

The food chains in the ocean begin primarily with the algae, which synthesize organic compounds from inorganic materials. In addition, anaerobic bacteria (i.e. bacteria which can live without oxygen) play a part, but this food chain has not yet been investigated in sufficient detail. The prerequisite condition for the assimilation process is that water, carbon dioxide, mineral substances and sunlight should be available. Since water and carbon dioxide are always present in sufficient quantities, the productivity of a region of the sea depends on the total calorie input due to radiation and on the mineral content. Part of the sunlight is reflected back from the water surface. With the sun at an angle of 5°, the quantity reflected back is 40 per cent, but when the sun is at an angle of 50° the reflection loss is only 3 per cent. In addition, the light is absorbed by the water and by the presence of substances causing turbidity; the different wavelength bands are differentially absorbed. Red and ultra-violet light penetrate less strongly than blue or green light, which is why the diver sees everything under water in a bluish illumination. The compensation point, at which assimilation and dissimilation (formation and breakdown of organic matter) are equal to each other, lies only a few centimetres below the surface of the water in turbid conditions, but in the very limpid Sargasso Sea it is situated at a depth of 100 metres (330 feet).

In places where ocean currents rise, the nitrates and phosphates resulting from decomposition of the organisms – these salts initially tend to sink – are brought up to the surface again. This favours a lush growth of algae, with the consequence that the fish population is dense. Such conditions are found, for instance, along the coast of Peru and along the west coast of Africa.

However, on average the sea is much less productive than the land. Precise measurements have shown that the average annual production of dry organic substance per square metre of open sea is about 100 grams (3 ounces a square yard), whereas in the coastal regions it is 200 grams (nearly six ounces) and in regions affected by upwellings (Peru) 600 grams (17½ ounces). Algae and seaweed raise this figure to 2,600 grams (4¾ pounds) while along the coral reef the figure is up to 4,900 grams of organic substance per square metre per year (9 pounds a square yard). These very productive zones are, however, relatively narrow and are consequently less important in proportion. The productivity of the land surface is generally considerably higher. Forests in the temperate zone produce 1,600 grams of organic substance per square metre per year (about 3 lb. a square yard), temperate grassland up to 3,200 grams, cultivated land (maize) 4,000 grams (7 lb. 5 oz.) and in the tropics (sugar-cane) up to 9,400 grams (17 lb.). Tropical rain forests produce up to 6,000 grams (11 lb.). The average productivity of the land is 1,200 grams of dry organic substance per year per square metre (2 lb. 3 oz. a square yard); as against this the sea, producing 120 to 150 grams of dry organic substance (around 3½ oz. a square yard), is decidedly unproductive. Although the larger area of the sea means that more sunlight falls on it than on the land, the productivity of the land surface is 25 to 50 per cent higher than that of the ocean. This is one major objection which must be presented to all those utopian dreamers who think that the sea is an inexhaustible reservoir of food from which thousands of millions of human beings will in the future be able to feed themselves. If the assimilating algae were spread out on the surface of the ocean, they would form a green film only one-quarter of a millimetre (roughly one-hundredth of an inch) thick.

Nevertheless, this film of algae is of increasing importance for our survival. We are in the process of denuding the Earth's surface by deforestation and by reckless exploitation of our cultivable land. The deserts are spreading, while at the same time our industrial installations and the internal combustion engine are giving off enormous quantities of carbon dioxide, with the result that in the last hundred years the carbon dioxide content of the atmosphere has risen by some 10 per cent. On the one hand this is encouraging plant growth, but if the figure continues to rise, it could lead to long-term changes of climate. Even now, the land area of the United States is regenerating only 60 per cent of the oxygen consumed. Thus the sea is to an increasing extent responsible for the regeneration and purification of our atmosphere. In spite of this, we are busily occupied in destroying this thin film of living matter upon which we are dependent, by turning the sea into a main sewer and a dump for poisonous waste. Some of these poisons, such as DDT, have an inhibiting effect on assimilation. DDT is selectively absorbed by the algae and is passed on in the food chain. The world-wide scale of this pollution is clear from the fact that the penguins in the Antarctic are already storing in their fatty tissues such large quantities of DDT that they would be unacceptable as food under our current food legislation. And this has happened in spite of the fact that DDT was only introduced on a large scale after the Second World War. Other poisons which we are thoughtlessly pouring into the sea can in fact constitute an even more dangerous time bomb. It is absolutely conceivable that our stupidity will one day lead to our destroying the life of all our oceans – and ourselves as well.

Ingenious specialists

The great ocean basins are relatively efficiently separated from each other by the continental masses. As a realm of life, the ocean is made up of a series of zones with very distinct living conditions. The coarser division distinguishes between the ecological associations of the open sea (the pelagial) and of the sea bottom (the benthal). These two regions require the creatures that populate them to adopt completely different adaptive techniques. The sea bottom offers both support and protection to the creatures which creep over it or adhere to it. These two facilities are absent in open water. Since protoplasm is always heavier than sea-water, all the inhabitants of the pelagial must somehow or other ensure that they do not sink down into the regions hostile to them. Algae need both light and warmth; in the cold depths of the sea they would perish. Consequently, parallel adaptations have led both animals and plants to develop a very wide range of devices as a protection against sinking: floats, which increase their buoyancy by increasing the surface area they present to the water, pockets containing oil or gas, the elimination of heavy skeletons and the like. In this entire development the organisms came to a number of parallel results: the small sea-slug

Glaucus, which floats back downwards on the surface of the sea, fills its gut with air. The violet snail *Ianthina* floats back downwards on a raft made of froth bubbles which it has itself produced, by stirring up air bubbles from the surface of the water, coating them with mucus and putting them together. The jellyfish-like colonies *Velella* and *Porpita* have developed their own gas chambers within their 'bells', and float on the surface by this means.

Thus these different organisms have independently solved the problem of maintaining themselves afloat by means which differ in detail, but are similar in principle. There are also other modes of adaptation to the environment which have been reached by parallel paths and which are shared by *Glaucus, Velella, Ianthina* and *Porpita* and by many other pelagic creatures: for instance they are all blue in colour, and have thus so adapted themselves to the colour of the sea that it is not very easy for predators to see them. Moreover, in their flight behaviour the inhabitants of the open water have developed certain specializations. The flying fishes and certain of the cuttlefish escape from their enemies by jumping out of the water and gliding for a certain distance, so causing their pursuer to lose sight of them. Fish which reside in the open sea often crowd together in schools. This too is a protective adaption against an enemy (see Chapter 7 for schooling behaviour).

The types of organism found on the sea bottom manifest entirely different characteristics. These denizens of the sea-bed can attach themselves firmly, creep into fissures or bury themselves in the sand and consequently develop heavy armour and skeletal structures or, if they live in the shelter of the rocks, can also bear brightly-hued signal colours. In addition, according to whether they live in the rocks or on the sandy bottom, other appropriate adaptations have been evolved.

The zones of the sea and their inhabitants

The ecological associations of the sea-bed and of the open sea can be further subdivided vertically. The surface regions are inhabited by a series of specialists. The siphonophores belong to the typical denizens of this region. It has also been discovered that the uppermost few millimetres below the water surface are populated by a specific world of small organisms which is called the neuston. The region of the sea bounded by the level to which light penetrates – down to some two hundred metres (660 feet) – is known as the epipelagic region. Down to the compensation point we speak of the euphotic zone. In the warm oceans the temperature falls off to about 20° C (68° F) at a depth of two hundred metres. The epipelagic zone is followed by the mesopelagic zone, which goes down to one thousand metres (3,300 feet). At the lower limit of this zone the temperature measured in warm seas is 10° C (50° F). Below this level extends the bathypelagic zone of the deep sea. Here the temperature falls to 4° C

(39° F), the lower boundary of this region being given as between 3,000 and 4,000 metres (some 10,000 to 13,000 feet). If we go still deeper into the abyssopelagic zone, the temperature drops to below 4° C. In the benthal, the uppermost zone we distinguish is the coastal zone or littoral. This comprises the entire continental shelf down to a depth of 200 metres (660 feet) and there ends as the continental slope begins. The coastal areas of the Earth cover some 28 million square kilometres (10.8 million square miles). To the zoologist the littoral is one of the most interesting regions, primarily because it requires of its inhabitants a series of extreme special adaptations. The uppermost section of this region, the tidal zone, is free of water once or twice a day. The creatures which live there are subjected to the constant impact of the waves – sometimes involving considerable forces – and this calls for special adaptive measures. The inhabitants of sandy waterline regions must be able to protect themselves against the abrasive action of the sand which is stirred up in the surf. Creatures which adhere to the rocks develop heavy armour, which resists the impact of the waves and protects them against drying-out; they also develop firm adhesive areas with the capacity to anchor themselves mechanically to the rocks – to cite only one or two of the special features evolved. Barnacles, limpets, winkles, edible mussels, oysters, the armoured sea urchins, starfish and the more agile crabs live here. Many of these species venture quite far inland, into the so-called spray zone, which is wetted only during storms or at the time of spring tides. Here we find barnacles, for instance, which can tolerate many hours of exposure to strong sunshine and can even resist weeks of dryness.

In the tidal pools left behind by the ebbing water, we find fish – mainly gobies – which have developed an astounding capacity for learning in their adaptation to this region. If the Atlantic goby, *Bathygobius soporator*, is suddenly frightened, it springs from pool to pool until it can get back into the sea. The American research worker Lester R. Aronson found that these gobies can cover a distance of up to 10 metres (33 feet) and could make use of as many as eleven pools. When he ran the water out of one of the pools, the fish jumped into the dry area. Consequently, it was concluded, they sprang blindly, moving only by memory. This observer noted that at high tide these fish learned the location of the hollows, and could recall this learned information over a period up to forty days.

The sub-littoral zone finishes at 200 metres (660 feet). Following on this is the bathyalic zone, the lower boundary of which is given as varying between 3,000 and 4,000 metres (some 10,000 to 13,000 feet). The area covered is ninety million square kilometres (34.8 million square miles). Below this again is the abyssal zone, covering 240 million square kilometres (92.6 million square miles). The deep sea trenches are known as the hadal zone. These features contain interesting local associations of organisms, since the trenches are isolated from one another.

The coral polyp

Of all these different habitats, the coral reef stands out by reason of the richness of species which it houses and the multiplicity of forms which it presents. Even rapidly leafing through the coloured plates in this book soon gives the reader an impression of the overwhelming wealth of different colours and shapes found along the reef. In the tropical oceans we can see forests of rock, peopled by shoals of jewel-like fish. Beautiful hedges of coral alternate with blocks of rock split by fissures inhabited by starfish and predators. The architects of this constantly changing landscape are the minute coral polyps. Their tubular body has only one opening, which is surrounded by tentacles. The closed end of the tube forms the base of the polyp and there creates as an exoskeletal structure a basal plate, which has an annular thickening around the edge. In addition, the basal plate is reinforced at six points by separate radial calcareous ribs. These ribs grow upwards, so that they and the basal plate which surrounds them finally extend as so-called septa into the stomach cavity of the polyp. By means of this exoskeletal structure, the polyp creates a shelter into which it can retire if danger threatens. Finally, if the septa have once reached a given height, the polyp creates a new basal plate above them. Whether, when this occurs, the lower part of the polyp is ligatured or whether it leaves its own refuge is not known. At all events, a continuous series of new levels is created and the stem of coral grows in height. Over thousands of years, this leads to the formation of huge reefs. Such reefs form the foundation of certain mountains in the European Alps.

The best conditions for rock corals are those in the warm seas, and it is only here that their activities lead to reef formation. As the water temperature increases, so also does the number of species. For instance, on the Barrier Reef on latitude 35° south, there is only one genus of coral. The temperature in this region varies between 10° and 25° C (50° to 77° F). In latitude 20° south, with a temperature of 20° to 30° C (68° to 86° F), we find forty genera and at 10° south, where the temperature is 24° to 32° C (75° to 90° F), there are sixty genera. In addition, the reef corals require sunlight, since their living tissues contain algae which live there symbiotically, using the sunlight to process carbon dioxide and the nitrogenous decomposition products of the polyps and produce oxygen.

This very varied coral landscape provides many organisms with the means of life. The coral in particular plays a decisive part in the life of these creatures, serving them as food, a place of refuge, a hiding-place and a home. Many of the creatures have become specialized in eating the coral polyps. The parrotfish bite off the shoots of the branching coral and gnaw at the coral blocks with their hard mandibles. Their function as sand producers along the reef is an important one. A whole series of butterflyfish also eat the small coral polyps. Among the invertebrates, the crown of thorns *(Acanthaster plancii)* has attracted considerable attention as a destroyer of coral. This large starfish climbs into the stems of coral and engulfs them with its stomach. The coral polyps are digested and all that is left behind is the white calcareous skeleton (Plate 14). In recent years the starfish have occurred in large numbers in certain parts of the Pacific, and have entirely laid waste individual reefs; this has also happened around Green Island on the Great Barrier Reef. It was feared that this epidemic could threaten the Barrier Reef and, as a result, represent a risk to the Australian continent, which – if robbed of its protective girdle of reefs – would be exposed to the destructive action of the waves. This is clearly an exaggeration. During a visit to the Barrier Reef in the summer of 1970, Hans Hass and I observed that the starfish were present in such masses only in certain clearly defined areas. There they were in the process of destroying the reef, but at the same time they were destroying the very basis of their existence. Once the corals have been consumed, the crown of thorns starfish disappear and the reef can go on growing.

Differing views have been put forward to explain these massive local increases. One of the suggestions made is that shell collectors have caused the damage by exterminating the snails which are the principal enemy of the crown of thorns. These epidemic occurrences can, however, be due to a perfectly normal phenomenon. It has been observed that many species undergo marked cyclical increase, even without human intervention, and in doing so destroy their own food basis, so that all but a few individuals die, following which the creatures which prey on them increase and a new cycle begins. The crown of thorns has, however, attracted attention and a very varied range of methods to destroy it has been tried out. One of these involves sending down divers to poison the starfish with injections of formalin. The tiny harlequin shrimp also came into the discussions about the crown of thorns in the last two years. Two German researchers, U. Seibt and W. Wickler, kept these shrimps in sea meadows to investigate their mating behaviour. In the process they discovered that the shrimps kill the crown of thorns. The shrimps break the hold of the starfish and eat away the soft parts. It may thus be possible to use this species of shrimp as a means of biological control of the starfish, should this prove to be necessary.

Fish that hide in the reef

As well as providing a source of food, the coral reefs offer hiding-places – and consequently protection from enemies – to a large number of sea-dwellers. The damselfishes of the genera *Chromis* and *Dascyllus* prefer to live among the coral branches. They feed in the open water, but if frightened immediately swim back to the refuge of the coral stems, and do not easily let themselves be driven out again. You can break the coral stem off and take it to the surface with the fish still in it. Even when the branch of coral is lifted out of the water, these small fish remain there.

If we take a closer look at a stem of coral, we shall discover many hidden inhabitants: sponges grow on most parts of the coral which have died off; brittle-stars and shrimps lodge between the branches and small fish *(Caracanthus)* hide there too. Many creatures have drilled into the coral stems, such as bivalve mollusks and small crustaceans. But all protection is relative. There are specialized fish which with their long tweezerlike snouts can even extract their food from these safe-seeming nooks. Checking over the population of a coral stem in this way soon reveals that it is a complete microcosm. This multiplicity of small habitats is the most striking characteristic of the coral reef. Each type of coral, each sponge and each small growth of algae contains its own little living society, each patch of sand has its own group of inhabitants which is typical of the particle size of the sand there; these are the things which make diving such an adventurous voyage of discovery.

These living-spaces for small life-forms combine to form larger typical undersea landscapes. Precisely as on land there are various types of landscape – fields, woods or hedges – with their typical societies of living creatures, we also find below the sea forests and hedges of coral, cliffs of coral blocks, screes, patches of sand and fields of algae. And each of these landscapes is inhabited by a characteristic group of creatures. Enormous numbers of fish of the most varied forms and colours swim around the hedges and blocks of coral. There are rectangular boxfish, rotund globefish, slender wrasse and high-backed angelfish. Most of these fish which are typical of the living denizens of the reef have one characteristic in common – they are exceedingly manoeuvrable. They can rapidly change direction, stop in one position, turn on the spot and skilfully steer between the branches of coral. In addition to this, many of them are astonishingly brightly coloured.

Colour as a signal

These striking colorations have generally been developed as a means of identification of the species. Just as the flags flown by ships indicate their nationality, the fish indicate the species to which they belong, and fishes of the same species recognize each other by their markings. However, not every one of these patterns – so striking to us – has this function. Many fish have a prominent dark stripe around their heads, passing over the eye. These are camouflage stripes which serve to hide the eye. Many of the sawtooth blennies, to which we have already referred, in fact attack the eyes of their prey – experiments by Wickler clearly showing that this camouflage is a protection for the eyes. Many fish have also developed a distracting eye spot below the dorsal fin, and this false eye attracts the attacking predator. In addition to this, however surprising it may seem, there are bright colour patterns which hide the fish. The jewfish *Epinephelus itajara* (Plate 71) and the small grouper *Cephalopholis miniatus* are covered all over with stippled spots. This stippling is striking to our eye, yet even we find it difficult at first to discern the fish clearly because of the disruptive coloration. The shape of the predator is to some extent masked, and this is the advantage of the markings. Many fish can also adapt themselves by changing colour and can even use this phenomenon as a means of 'communication'.

Adaptation to sand

The sand patches house a very different association. At first sight these sandy areas seem very sparsely populated, and even desert-like. Only when we look more closely do we see depressions and holes in which a very varied range of creatures is hiding. Since the surface of the sand offers no protection, most of the sand-dwellers have developed the capability of digging into the sand. Many of them, like the various kinds of sea-urchin and snail, burrow in the sand as long as they live. Others again live in tubes. There are worms which extend from their dwelling-holes long, sticky tentacles which spread out in all directions over the sand surface. Particles of food remain adhering to these tentacles and from time to time the worm gathers its catch. One type of worm lives in a U-shaped tube and uses fin-like appendages to create a current of water in which it spreads a net of mucus, from which it periodically licks off the particles trapped.

While the sand is inhabited by a large number of creatures, the surface of the sand is sparsely populated. It is here that we find starfish, sea-cucumbers, prosobranch snails and a series of sand-coloured fish. Many of these are flat like the soles and plaice. Others construct tubes which they line internally with tiny stones after the fashion of a well. I have always been particularly interested in these sand adaptations, and have therefore studied them closely around the Nicobar and Maldive Islands. Especially on rainy days, when it was not worth making excursions into the reef, we examined the flat sandy or muddy bottom around our anchorage. Over and over again surprises awaited us. Once we observed a whole regiment of individual corals about half an inch across, advancing across the surface of the sand; closer investigation showed that they were living in association with worms which pushed them forward, so preventing the tiny corals from being buried by the sand. In return each worm had a partner which protected it, since it lived in a tube in the limy skeleton of the coral. On another occasion I came upon a whole group of sea-urchins. Each was accompanied by a swarm of cardinalfish. When danger threatened the fish hid between the spines of the sea-urchins, while the cardinals – as it were in payment – cleaned the spines. Another defensive-and-offensive alliance which we discovered showed gobies living in association with a shrimp. The shrimp dug away the sand like an excavator shovel, creating a hole which both could use. In return the gobies gave warning of the approach of danger. But the most remarkable sand adaptation which we discovered was that of the garden eels. When I saw them for the first time, I thought that I was looking at a field of algae. Only when I looked more closely and saw the fish slowly withdrawing tail-first into their tubes, did I realize that I was looking at a genuine meadow of eels. Each eel had inserted its tail into a tube. So placed, it extended its head into the open water above, where it could catch its prey.

Typical reef formations

There are various types of coral reef formation. Around islands and continents the coral forms fringing reefs. If the coast slopes away steeply, the coral can then luxuriate even near the shore. At flat places along the coast however the shallow area is sandy or cemented together by calcareous algae to form a plate of rock. The strong movements of the water in these areas allow only the lumpy and compact types of coral to prosper. Frequently, the shallow region ends abruptly at a wall of calcareous algae, beyond which a steep slope clothed with coral falls off into the deep water. This wall of calcareous algae is broken by channels through which the water flows out of the region of shallow water at each change of tide. The precipitous wall of the reef is covered with a lush growth of coral for the first 20 metres (65 feet), and from this point on the coral growth falls off rapidly. At a depth of 40 metres (130 feet) we generally find only isolated coral stocks in a waste heap consisting of coral debris.

As an island subsides in the course of time, the living coral reef moves further away from the coast, since the coral grows most freely on the seaward side. In this way the reef finally becomes a barrier of rock lying just off the coast (Barrier Reef, Plate 2). If the island sinks altogether, this barrier reef remains as a ring-shaped wall, giving rise to the formation known as an atoll. This ring of coral encloses a lagoon, which can be more than 40 metres deep (130 feet) in the case of large atolls (Plate 1). The bottom of the lagoon is covered with sand and ooze, but coral blocks can also grow on the bottom. The atoll is interrupted in certain places by surge channels through which the water in the lagoon flows in and out at each change of tide. Storms throw up coral waste onto the reef, and this leads to the formation of small islands which ultimately become populated by land organisms.

In the Maldive Islands we were able to investigate islands of this sort in all stages of formation. One in particular is etched in my memory. The free area of land at high tide was some 10 m × 7 m (33 × 23 feet). This tiny islet lay almost invisible in the blue water. The coral debris was sharp-edged and dark in colour, as though burned by the sun, and heaped up like a wall around the island were coconut shells, driftwood and the horny skeletons of lorgonia. Otherwise the islet was completely bare, but a single small bush grew in the middle, the advance guard of the land flora. I examined this bush closely and found that already a few caterpillars had begun to eat its leaves. On the stem was a small beetle and in the half-rotten leafage at the foot were a couple of spiders. That bush was a small island on the islet – an island of land life. Many times this little bush provided me with shade while I rested after the long return to the surface from a deep dive.

The caves of the Maldives

In an atoll we can distinguish between a seaward outer reef and an inner reef on the lagoon side. The inner reef is often a pleasant spot, since there are no strong water currents and in addition we often find an extensive sand area towards the shore. The conditions along the outer reef have been described above. In the Maldive Islands, the outer reefs frequently form steep cliffs penetrating into the depths. Forty metres (130 feet) down we repeatedly discovered spacious caves. These are also found in other regions, e.g. in the Caribbean. They are probably wave-cut cavities formed in the recent Ice Age. At that time the level of the sea was many feet lower than now, since large quantities of water were locked up in the ice-caps. These caves in the Maldives are most impressive. As the diver enters with a lamp he beholds a magnificent, brightly coloured world. The walls are clothed with yellow, red and violet sponges. Venus fans, a form of horny coral, together with sea-lilies and pale soft corals, are set around the edge of the cave, while in the interior there are often large shoals of red soldierfish, with fat groupers between them. And in the narrow fissures are dozens and dozens of delicious lobsters.

In the atolls of the Maldives, mushrooms of coral frequently grow to considerable heights from the bottom of the lagoons. Once they reach the surface, they begin to spread laterally, and as they increase in extent the centre dies out. The corals perish and are converted into sand, and atoll-like formations are produced. This revealed to us that atolls are formed not only on the fringing reefs of sinking islands, but also on islands in full growth. Hass investigated this type of atoll formation and, as far as I know, was the first to describe it.

Pioneers of ocean research

Leafing through the illustrations for this book it struck me what astonishing progress has been made by underwater photography and by diving technique in the last twenty years. What is more, it is not so very long ago that the tropical oceans were considered as unsafe for unprotected skin-divers because of the numerous sharks. Hans Hass was the first to breach this rule, when in 1939 he dived in the Caribbean off Bonaire and Curaçao in the company of Alfred Wurzian and Jörg Böhler. There he learned that the thing to do was not to swim away from the sharks, because this triggered their prey-catching reflexes. The diver must wait for the shark and if need be scare him off with a jab from the harpoon, although generally the rapid movement itself is sufficient. The diver then swims towards the shark, which is easily put to flight. Other species react to shouts under water. We have made use of these experiences in all our diving expeditions and have only rarely had to take to flight ourselves. In my book *Land of a Thousand Atolls* I gave exhaustive descriptions of our experiments with sharks. I share Hass's view that the shark is the sea-dweller with the most finely-developed body shape. Every movement of these creatures embodies great power combined with unsurpassable grace.

Hans Hass brought a large number of fine underwater photographs back with him from the Caribbean, and many of them are striking behavioural records, such as the sensational series depicting the trumpetfish which rides on the back of big fish, swimming with them and so using them as stalking-horses to approach its smaller prey (Plate 11).

The techniques of diving and of underwater photography rapidly developed under the initiative of Hans Hass, and, independently, of Jacques-Yves Cousteau in France. A decisive day in the history of skin diving was 12 July 1942, when Hans Hass became the first fish-man by diving off the Greek island of Ari Ronisi with skin-diving equipment, using an oxygen apparatus manufactured by the Draeger Company. In the following year, Cousteau experimented with a skin-diving apparatus in which compressed air was used instead of oxygen. This was found later to be more reliable and conquered the world as the 'Aqualung'. Hans introduced this new technique as a method of carrying out scientific work, and nowadays all over the world, scientists of the most varied disciplines employ diving in their researches.

The new technique has had a decisive effect on various aspects of marine research, but Hans Hass was responsible not only for advances in diving technique and underwater photography: his research also attracted many young people to underwater hunting. Whereas he went after fish with simple hand-held harpoons, the modern underwater hunter frequently employs compressed-air-, elastic- and spring-powered harpoons, and the fish are left with very little chance of escaping. It does not demand much skill to bring down a fish by means of a compressed-air harpoon. Hass was horrified when, after a lapse of some years, he visited areas where he had himself previously hunted. He found himself swimming through deserted reefs.

Coral reefs in need of rescue

The sea is not inexhaustible. Large fish, such as groupers, are found over reefs only at considerable distances from one another, and systematic hunting very quickly exterminates the large fish of a reef. Today, so many fish have been shot that large coastal areas of the Mediterranean have become rock deserts, and in many tropical reefs the length of the only fish still to be found does not exceed a handsbreadth. Hans Hass, who feels himself one of the persons responsible for this state of affairs, has already published a manifesto against underwater hunting. In addition, Cousteau's appeals to keep the oceans pure and to protect the wild creatures that live in them have received worldwide attention.

Every type of mechanical underwater weapon should be completely banned. This, you will say, is impossible to achieve. Such a step would in the first instance mean a considerable loss of revenue to the manufacturers. But we have to bear in mind much more is at stake than that. Diving as a sport loses much of its attraction if the stocks of fish are destroyed. This in turn means that fewer underwater cameras, sets of diving gear and similar equipment will be bought. In addition, underwater hunting as a sport with a hand spear should continue to be allowed. It is a difficult sport and the fish have an excellent chance of making good their escape. A little while ago I attended a meeting of sporting divers who made a demonstration by laying aside their mechanical harpoons. This is an example which should be followed, so that our children and grandchildren can also see the exciting beauty of the underwater world. Our task is to further understanding of the living relationships in the sea, since we are always ready to preserve what we can understand and admire.

Such is the purpose of this book.

Irenäus Eibl-Eibesfeldt

16 *(Opposite)* Reef zones.
Long fringing reefs extend along the coast of the Red Sea, as is the case here on the Sinai Peninsula. Generally they comprise the tidal zone, the sandy lagoons, the broad top of the reef and the steep incline down to the open water. All the living creatures found here have developed close adaptations in body structure and in behaviour to the conditions in these zones.

◁17 On this beach zone in the Red Sea there are settlements of sand cones thrown up by the male ghost crab *(Ocypode saratan)*. These sand pyramids are social signals to others of the species and are used by their owners as a means of orientation.

18 The male ghost crab carries the sand out of a spiral tunnel and throws it onto the top of the pyramid. The crabs' building activity begins in the spring at the start of the breeding period, since the pyramids are intended as a mating signal to all the females. At the same time they serve as 'defence-threat signals frozen into a solid form', used to frighten off other males.

19 The stalked eyes of the ghost crab can be folded away sideways into small depressions in the crab's carapace.

18

19

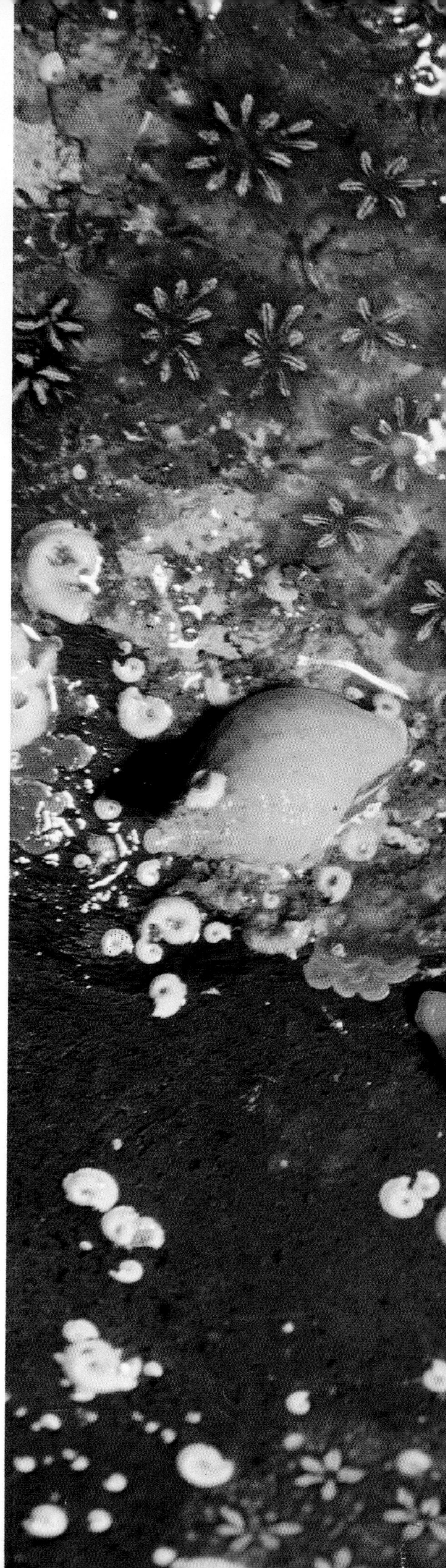

20 Inhabitants of the spray and tidal zones. The brittle-star *(Ophiocoma scolopendrina)* is one of the first species to move landward and settles in pools in the tidal region. By means of its long, mobile arms it plucks algae from the bottom or collects the small particles of food floating on the surface of the water.

◁21 In depressions in the rock the *Chiton* settles into place, its body being the precise shape to fit snugly into the holes. It attaches itself by means of suction, to ensure that it is not dislodged by the water at high tide.

◁22 *Planaxis sulcatus,* a snail, occurs in the Red Sea. These snails mass together in conspicuous groups, as a means of preventing themselves from being carried away in the surf at high tide and as a protection against drying-out at low tide – an adaptation to ensure survival in the tidal regions.

23 Pelagic medusae drifting by accident from the open sea into the intertidal zone where they die near the coast.

24 The rocks of the tidal zone are peopled by colonies of tightly-adhering tunicates *(Botryllus),* the gelatinous surface of whose bodies protects them against becoming dried out at low tide.

◁ 25 The upper ridge of the reef stands clear at low water. It drops off steeply into deeper water on the seaward side. On the landward side the corals die off slowly, break down and eventually form sand.

26 Sea-horses *(Hippocampus),* inhabitants of the algal meadows, hold tight by means of their mobile tails. They are a curiosity among fish, since it is the male which bears the young.

27 A typical inhabitant of the top of the reef is the surgeonfish *(Acanthurus sohal),* which occurs only in the Red Sea.

28 In flat sandy lagoons or in the sand and mud area, live the sea-cucumbers *(Holothuria),* which are related to the sea-urchins and starfish.

29 The giant clam *(Tridacna)* closes the two valves of its shell as soon as a shadow falls on it – an adaptation against predators. On the Great Barrier Reef, these clams can reach a shell length of 1.5 metres (5 feet) and a weight of 200 kilograms (440 pounds).

26

27

28

29

31

32

33

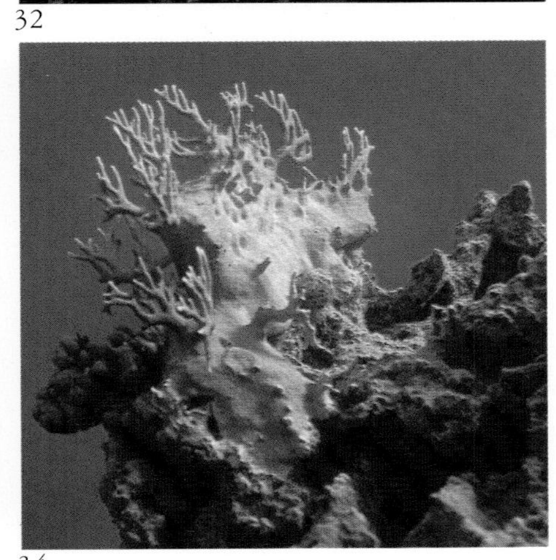

34

30 *(Opposite)* Beyond the reef ridge the reef itself plunges steeply. In the upper regions, there is a particularly lush growth of coral. A typical denizen of this region is *Anthias.*

31 The blenny (Blennidae family) skulks in small fissures in the rock, where its coloration serves as good camouflage. Other species inhabit the empty tubes of polychaete worms or nest in cracks in the rock.

32 The purple nudibranch (Aeolidiaceae family) presumably frightens off predators by its striking coloration. Some species even eat the stinging cells from coelenterates and defend themselves with these stolen weapons.

33 Starfish – in this illustration a *Protoreaster* – also occur in the various regions of the reef. Many of them are predators and often attack bivalves. While many species are photophobic and shun the light by hiding, others live in brightly-illuminated zones.

35

34 *(Preceding page)* Fire coral *(Millepora)* grows freely on the steep seaward side of the reef. This is not a true coral, but belongs to the colony-forming hydrozoans. Contact with these pseudo-corals produces considerable pain and can lead to severe nettle-rash.

35 *(Preceding page)* The brain corals – in this illustration *Platygyra lamellina* – frequently have their body surface very much enlarged by convolutions to increase rhe amount of metabolically active tissue. A larger surface area means that there will be a more rapid deposition of lime and thus that growth, too, will be more rapid.

◁ 36 A fan coral of the Caribbean Sea colonized by a hydrozoan polyp *(Pennaria)*. The polyps, both of the coral and of the hydrozoan, feed on plankton, which they trap with their tentacles. The concentration of plankton is at its richest along the reef at night rather than in the daytime, hence most of the invertebrate plankton-feeders have become nocturnal.

37 The polyps of this soft coral *(Sarcophyton ehrenbergii)* from the Red Sea generally extend their tentacles into the water at night. These tentacles immediately retract into the body of the coral the moment they are touched. This rapid contraction protects the coral polyps against predators.

38

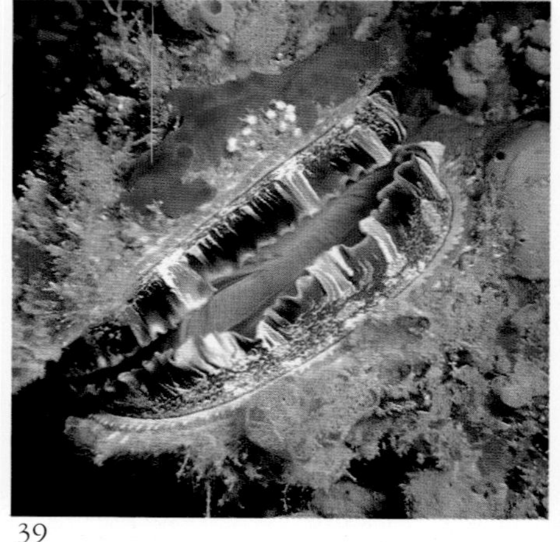

39

40

41

42

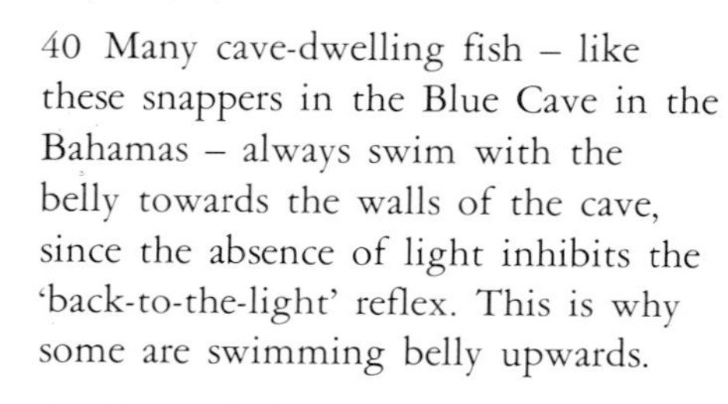

38 The soldierfish *(Holocentrus rufus)* lives in the fissures and caverns of the steep reef walls. It can produce sounds which can even be perceived by divers under water.

39 The thorny oyster *(Spondylus* sp.*)* lives in clefts and, like the giant clam, shuts its valves the moment a shadow falls on it.

40 Many cave-dwelling fish – like these snappers in the Blue Cave in the Bahamas – always swim with the belly towards the walls of the cave, since the absence of light inhibits the 'back-to-the-light' reflex. This is why some are swimming belly upwards.

41 Deep, dark caves in the reef wall are inhabited by spiny lobsters *(Panulirus)*. Only at night do they leave the safety of their refuge, and then they even crawl up as far as the reef ridge.

42 Sponges, too, are inhabitants of the dark zone. They are often colonized by other creatures – here by nudibranchs. The interior of the sponges, frequently divided into a number of chambers, offers excellent accommodation for colonists.

43 Many reef landscapes, like this underwater scene in the Bahamas, are formed by fan, horn and gorgonian corals. The extraordinary variety of the coral world offers food and refuge to a very wide range of species.

43

45

46

◁ 44 The garden eel *(Gorgasia sillneri)* in the Red Sea settles in sandy bottom areas swept by currents. These creatures fish for plankton and are adapted in both body structure and behaviour to their mode of life.

45 These pyramids – up to 1 metre (40 inches) in height – on sandy or mud-covered bottom areas are connected by a system of tunnels to a funnel-shaped depression. It was not possible to catch the creature inhabiting this system of channels.

46 The Ribbontail ray *(Dasyatis lymma)* lives on the sandy bottom at the foot of the reef. For protection against predators, many inhabitants of the sandy zone have developed a special flight technique, while others avoid casting a conspicuous shadow, by adopting a low body profile; in addition they wear camouflage patterns.

47 *(Overleaf)* The sand-diver *(Gunellichthys monostigma)* makes a tube in the sand into which it flees headfirst when threatened. While eating, it remains stationary in the current above the bottom and fishes for plankton.

1 The zones of the reef and their inhabitants

Diving gear on my back, I am swimming down a reef wall on the coast of the Red Sea. Although I have done the same thing many times before, I am always surprised and fascinated anew by the tremendous variety and beauty of the landscape spread out before me. I glide over undersea cities with stone gardens, slender minarets, round domes and bizarre towers. Fantastic shapes surround me – tremendous mushrooms, huge bulbous growths, and elegant fernlike fans. Grotesque forms flit through this magical world of Nature's making, their bright colouring made still more striking by the greeny-blue background against which they move. It is quite impossible to pick out individual fish from this teeming multitude. Again and again fresh individuals swim into my field of view and disappear into the fissures in the reef.

The basis of all this busy colourful life is constituted by the coral structures which have been formed as the result of the gigantic collective activity of minute polyps. Over periods of thousands of years these tiny architects of the sea construct atoll rings, broad barrier reefs or long coast-fringing reefs, as here in the Red Sea. It was the coral polyps that built the Great Barrier Reef – 2,000 kilometres long, 150 kilometres wide (1,240 × 95 miles) and with a steep precipice plunging into the open sea to a depth of almost 2,000 metres (6,600 feet). How could these animalculae possibly have created so gigantic and varied a landscape below the sea?

Each piece of coral consists of a dead calcareous skeleton, coated with a layer of living material – the coral polyps. The polyps continuously secrete lime and thus build up the coral stock. However, this is true only of the stone corals (madrepores); the many multi-coloured soft corals, which are also an essential part of the reef society, cannot produce a limy skeleton.

The coral polyp consists of a multi-cellular tube, closed at one end by a basal plate, and at the other by a mouth disc. This latter is surrounded by a large number of small tentacles, which the polyp uses to trap plankton.

The polyp, which is a creature of very simple organization, possesses no specialized breathing, circulatory or excretory organs. Its food is digested in the body cavity, which is divided into a number of chambers, and is then distributed and excreted again through the mouth.

The living conditions of the polyp

In spite of its primitive structure, the coral polyp is an organism with very specific requirements. In order to grow, it requires clean, warm water at a temperature of at least 20° C (68° F). In addition it requires ample supplies of oxygen, which it finds primarily in shallow water in constant movement. If a flow of fresh water carries dissolved substances and dirt into the seawater, the coral stops growing or dies off. Changes in temperature, lack of light or the slow deposition of small particles can also cause the death of the polyps. Most polyps are self-cleaning and remove by means of small cilia all the foreign particles which come into contact with their body surface. When a tarred road was built in 1969 along the coast of the Sinai Peninsula in the Gulf of Aqaba, causing large quantities of dust, sand and other polluting materials to enter the sea, many corals stopped growing after some time or died. The self-cleaning capacity of the polyps was inadequate to cope with such a flow of impurities.

Formation of the calcareous skeleton

The coral polyp secretes the lime (calcium carbonate) on the external surfaces of its basal plates. The calcium is present as soluble ions in the sea-water and scientists have used radioactively labelled calcium to prove that this is the source from which the polyp extracts the calcium. However, we do not yet know precisely how it produces solid crystalline lime from the soluble ions it takes in.

The rate of lime deposition varies with the different species of coral and is markedly dependent on the illumination. According to the American ecologist T. F. Goreau, this process takes place most rapidly in the middle of the day in unimpeded sunlight, dropping by half that rate if the sky clouds over, or by 90 per cent in darkness.

Corals are able to produce daily some 10 grams of lime per square metre of polyp surface (0.3 ounce per square yard). This requires considerable quantities of energy, which must be provided by the metabolic processes of the polyps. Scientists investigating coral have attempted to determine the metabolic rates of the various species of coral. The measurable variable used as an index was the polyps' consumption of oxygen. Results show that the intensity of metabolism of the polyp is greater than that of a human being at rest. In one hour a man requires about 8 milligrams of oxygen for each gram of body weight, while the average figure for the coral is around 20 milligrams. Thus the structurally simple coral polyp has a far higher requirement than that of a highly-organized vertebrate. This clearly illustrates the considerable quantities of energy which the polyps must transform in building coral.

The source of energy for the metabolic process

For a long time it was believed that the coral polyp obtains its metabolic energy from processing the plankton on which it feeds. Its tentacles are provided with complicated trapping and 'shooting' mechanisms specially adapted to trap plankton; small, spirally-wound arrows are very rapidly expelled from the nematocysts (nettle cells), reacting to the slightest touch like a coiled spring; these missiles penetrate into the body of the victim; at the same time there is released a poison which paralyses the prey. Since plankton is particularly plentiful at night, most types of coral polyp have become nocturnal in habit.

Recent experiments have shown that the polyp does not feed solely on plankton and that certain types of coral actually dispense with it altogether. During these experiments, when all the plankton was filtered out of sea-water, some species of coral nevertheless continued to grow. We know nowadays that the growth of these corals is dependent on single-celled algae which live in the soft parts of the polyp's body. These algae belong to the class Dinophyceae, which manufacture oxygen and carbohydrates by photosynthesis under the action of sunlight, and furnish these substances to the polyps. In doing so, the algae also process the carbon dioxide produced by the breathing of the polyps and break down nitrogenous products of metabolism. Without sunlight these symbiotic algae could not exist. When corals were experimentally maintained in darkened aquaria, all the species symbiotically associated with green algae ceased to grow, while the species of coral having no such association continued growing.

In recent decades many research workers investigating coral have reached strongly conflicting views about the significance of the algae in the polyp's feeding habits. The question which arises is why corals should in fact have developed such complicated plankton-trapping mechanisms if they do not of necessity require plankton to live on. There are indications that corals do not obtain the phosphorus which is vital for their metabolic processes in solution from the sea-water, but can in fact procure their requirements from small quantities of animal plankton. (A distinction is drawn between zooplankton or animal plankton, and phytoplankton – which is plant plankton.) This would bring us back to the conclusion that the plankton is indispensable and is one of the determinative factors in the survival of the coral. Research into this interesting field of polyp metabolism is still being actively pursued and it will only be when we have a clearer understanding of the metabolic processes involved that the significance of the plankton and the algae in the feeding cycle of the polyps can be correctly assessed. One thing, however, is certain: without the aid of the green algae, no coral reefs would have been formed.

The birth of a new coral stock

Let us now follow in sequence the stages in the life of a coral polyp. Polyps can propagate themselves either asexually by cell division, or sexually by fusion of male and female germ cells. A few corals are hermaphrodites, being of both sexes at once.

The ripe egg cells are fertilized either within or outside the body. Small free-swimming larvae hatch out of the fertilized eggs and are transported over wide areas by the water currents. Once they find a suitable support – generally an existing reef – they establish themselves there and rapidly change into small polyps. No one has yet investigated the question of how the larvae recognize a suitable substrate.

The young polyp now grows rapidly. It secretes lime intensively and surrounds itself with a cup-shaped wall. It can withdraw into this exoskeleton as a protection against its numerous enemies – fish and invertebrates. Eibl-Eibesfeldt has already shown in his introductory chapter that the pedestal of the polyp is reinforced at six points by separate radial bands of lime and that these bands or septa extend into the large internal stomach cavity. Once the septa have reached a certain height, the polyp begins to secrete a new foot, thus producing a further storey. The fully-grown polyp now divides asexually; new polyps are produced in the immediate vicinity and these too can multiply asexually. All these polyps ultimately form the coral stock, which continues to grow upwards. There are, however, certain species of coral, such as the *Fungia* or mushroom coral which occurs widely in the Indo-Pacific region, in which the building is done by only a single individual.

At certain times the polyps again form sexual cells and swarms of small larvae are distributed all over the reef; thus the cycle between sexual and asexual propagation recommences.

Size and growth of the corals

A coral does not continue its growth indefinitely, but has a particular maximum growth. For instance, the hemispherical brain corals have a diameter of 2 metres (6.5 feet), the large madrepore plates 3 metres (almost 10 feet), whereas the gorgonian fans *(Lophogorgia)* can reach a height of 3 to 5 metres (10 to 16 feet) and the massive *Porites* coral stocks as much as 6–8 metres (20–30 feet) in height.

Sunken wrecks offer ideal possibilities for establishing the approximate rate of growth of corals, provided that the date on which the ship sank is known. *Acropora* corals grow 10–25 centimetres (4–10 inches) per year; the bulky *Porites* and brain corals, on the other hand, grow at only 1 centimetre (0.4 inch) of diameter per year. Measurements of the increase in weight of a coral stock showed that the massive corals have the lowest annual rate of increase. The larger the surface area of the coral, or the looser and more porous the skeleton, the greater the rate of growth. Measurements showed that *Pocillopora* and *Porites* species have an annual rate of increase of weight of 100–200 per cent, many species of *Acropora* 400 per cent, while species of *Montipora* have a rate as high as 1,200 per cent. A species of coral will establish itself the more successfully on the reef (compared with other species), the higher its rate of secretion of lime and the more intensive its metabolic activity, in other words, the more efficient its polyp colonies are. In addition, when the coral has a large surface area, it can secrete more lime. The convolutions of a brain coral or the multi-branched structure of an *Acropora* thus serve to provide the species with a larger metabolically active surface, so achieving a higher rate of growth. The species of coral which are most widely distributed over the reef are those which have evolved most recently; their efficient equipment enables them to gain an advantage in the struggle for survival. However, it remains true that certain risks are attached to such specializations. If the ecological conditions are unfavourable, it is precisely these younger species of coral which die off first. Individual polyps whose metabolism is less intensive are less demanding and have a significantly higher degree of resistance. This is why they also occur in the regions inside the reef, which can no longer be colonized by the fast-growing coral species.

It is only very recently that American scientists have observed how polyps feed on other species of coral which they find in their immediate vicinity. After making contact with their tentacles, they extrude filaments containing digestive juices, spreading these filaments over their prey and gradually consuming it. In this way clearly visible boundaries are drawn at the point of contact between the polyps. This behaviour has been labelled 'aggression'. In certain coral reefs off Panama, the fast-growing species of coral are less 'aggressive' than the slower-growing types.

The formation of the reef zones

The differing requirements in living conditions, peculiarities of metabolism and behaviour of the polyps lead to segregation of the species within the reef, so that the living space is divided into a series of zones. Let us assume that a few corals have been able to settle in the vicinity of the coast. The first event is the formation of a narrow band of coral running parallel to the beach. Other colonists arrive: algae, jellyfish, sponges, snails, bivalves etc. The surf brings in the plankton and oxygen which are indispensable to enable the coral to grow further.

In the course of time, this band becomes wider and forms a horizontal ridge of reef, continually growing outwards towards the open sea. The species of coral which have the highest metabolic activity are located along the leading edge of the reef, since they must have optimum living conditions to maintain their intensive growth, and consequently require ample quantities of light, oxygen and plankton.

On the coast side, the ridge of the reef slowly dies out, since the conditions obtaining there gradually become less and less favourable for the corals. The sediments stirred up by the action of the surf are deposited in the areas of the reef ridge near the beach; the movement of the water has now become sluggish and exchange is inefficient, so that the quantities of oxygen available are no longer adequate. The dead corals leave behind calcareous debris, which is slowly ground down to form sand and becomes overgrown with algae.

The areas near the coast are regularly exposed and dried out at low tide. Depressions in the rock contain tidal pools which are subject to wide variations in temperature and salinity.

Between the old reef ridge and the rocky tidal zone, sandy lagoons often form and scattered new growths of coral occur here. The boundary between the land and the sea finally becomes no more than a spray zone, which is wetted only by wave spray at very high tides.

Let us look again at the edge of the reef. The nearer we get on the ridge of the reef to the open sea, the more living corals we find. Right at the leading edge, where the surf breaks incessantly, there is a particularly lush growth of coral and here too the variety of fish species is particularly great and colourful. Along this line the ridge of the reef drops steeply into the water. The upper areas of this relatively steep precipice exhibit equally rich colonies of coral, and numerous fissures, cavities and small crevices offer refuge to a large number of organisms. The deeper we dive, the weaker the light. The movements of the water become less and less marked and the supply of oxygen falls off. As a consequence of all this, coral growth becomes progressively reduced. Generally the only species to grow in these deeper zones are brightly-coloured gorgonian fans, soft and fan corals, which require less light and oxygen.

At fairly great depths, say from 40 metres (130 feet) on, there is no further coral growth at all. At this point there is generally a heap of the debris of dead coral at the foot of the reef, which slowly gives place to extensive deposits of sand or mud. The fish and other marine creatures have adapted to the various reef zones in the most varied manners possible. Each zone has its own associations and exhibits special adaptations in body structure and behaviour which are typical of the denizens of that zone.

Specialization as a means of colonization

Entering a coral reef from the land side, we generally cross a beach zone of fine coral sand, partly mixed with bivalves and dead coral branches. Certain creatures which have adapted themselves to the special ecological conditions of beach life inhabit these flat regions where there is no cover.

Off the coasts of the Red Sea and the Indian Ocean, the coral sand is often thrown up in small pointed pyramids (Plates 17, 18). It was a long time before the secret of these pyramids was discovered. These pyramids of sand are built by the males of the ghost crab *(Ocypode).* The male uses his right or left claw to dig out a spiral cavity in the sand. The waste sand is thrown out and built up in pyramids, the crab meanwhile levelling the path to the top of the pyramid. The pyramid is always situated along an extension of the line of the underground tunnel and thus marks the entrance to it. In this way the crabs can always find their way back to their own homes. The German behavioural scientist K. E. Linsenmair also discovered that these pyramids have an important social function. On the one hand they serve as focal points for the establishment of new colonies, which can comprise as many as 2,300 pyramids. New arrivals observe a minimum distance of 134 centimetres (4½ feet). In this way the colony gradually spreads outwards in all directions. For rival neighbouring males these sand pillars are fear-triggering signals and invaders are attacked savagely and driven off.

In addition to this, the pyramids serve as mating signals which attract the receptive females and show them the way to the mating place, which is in the spiral excavation. The males have largely lost their courtship display and seem to rely entirely on the attractive power of their sand pyramids. Only when the female appears at the entry to the excavation does the male emit an acoustic signal, to indicate to the female that the occupier is present. The excavated cavity is thus at one and the same time a dwelling- and a mating-place.

The ghost crabs leave their excavations for feeding, running along the edge of the water where they can pick up many little particles of food in the strip washed by the sea. If danger threatens they retire quickly to their holes. Once again the sand pyramids serve to orientate them.

We can frequently find in the wet sand the ring-walls of the ball-crab *(Dotilla sulcata)*. This crab sieves through the sand to find particles of food and lays out the waste remaining after use, in the form of balls distributed in a particular ring-shaped pattern around its dwelling-hole. In doing so it leaves tracks clear between the rings, which it follows when it needs to run quickly back into its hole for refuge. There are dangers on every hand on the flat expanse of bare sand, where birds and lizards constitute a particular threat to the crabs.

Other *Dotilla* crabs seek their food as they walk in huge groups along the edge of the sea. The individuals do not have their own dwelling-tubes, to which they can flee if threatened. These crabs have an entirely different adaptation to enable them to escape from their enemies – they simply dig themselves in with their claws. The beach-dwelling hermit crab *(Coenobita)* has found an even more effective means of escape from its foes; it hides in a hard, empty snail shell, closing the entrance by means of its claws, which precisely fit the shape of the shell. What is more, living in the snail shell enables it to avoid drying out of its body under the effects of strong sunshine. Such are the protective measures which enable the creatures of the bare, hot beach zones to survive.

The spray and tidal zones

The spray zone constitutes the boundary strip between the land and the sea. It is in this region that the variations in living conditions are particularly marked; nonetheless, a large number of creatures have established themselves in this area which is bleached by the sunlight and washed by the sea. In depressions in the rocky substratum are the *Chiton*; coat-of-mail shells, in body construction these creatures exhibit complete adaptation to their surroundings. The flattened body, compressed in its hole, exerts a strong hold and can be extracted only with great difficulty. Thus even heavy wave movements do not dislodge them, and they are in addition protected against enemies and against drying out in the hot sun. Limpets of the genus *Patella* are found in the spray zone and in the tidal zone. By moving their pyramidal shells they can obtain a precise fit with the support on which they sit. *Patella* makes grazing forays in the immediate vicinity of its home but in an astonishing manner returns each time exactly to the point from which it started. This it succeeds in doing by marking its site by means of specific marker substances from its own body and also by 'laying a trail' of this substance on the track it follows. Each limpet has its own perfume code, so that there is no likelihood of there being any confusion with other individuals. All the limpet has to do is follow its own scent and, like a railway locomotive on its rails, it returns to its own home. This facility for always finding precisely the same place – termed homing behaviour – increases the individual's chances of survival.

In the tidal zone, which is immediately below the spray zone, the plants and creatures are exposed to even wider extremes of conditions as a result of the periodic alternation between high and low tides. Periods in which they are out of the water are followed by periods when they are submerged, so that the organisms are obliged to revert to an aquatic pattern of life. The tidal pools in the ebb zone rapidly heat up under the hot rays of the tropical sun, the water evaporates and the salt content is raised. In order to survive, the inhabitants of the tidal zone must have the highest possible resistance to variations in temperature and salinity, and indeed have developed complicated mechanisms to maintain themselves alive in this zone.

Snails of the species *Planaxis sulcatus* (Plate 22) crowd close together in large numbers. At low tide this pattern of behaviour protects them from being dried out and at high tide offers a means of resisting the undertow.

On the island of Nosy Be, Madagascar, I once observed a drama of nature, of a kind which cannot be infrequent in the tidal zone. At the spring tide a large shoal of small sprats *(Spratelloides gracilis)* had by chance entered a tidal pool and had left it too late to return to the open water. The shallow pools had heated up rapidly during the daytime, with a resulting increase in salt content. Certain death was all that these fish could have expected when I rescued them by the bucketful from their accidental grave.

Deeper tidal pools offer homes for the young of butterflyfish, surgeonfish and domino damselfish. Moray eels may also remain in tidal pools in order to survive during the period of low tide. Only when the tide is higher do they come out again and go hunting for food.

One creature which has adapted particularly well to the tidal zone conditions is the amphibian mud-skipper *Periophthalmus* – a fish which lives a semi-aquatic, semiterrestrial life. Its pectoral fins also serve as small legs to enable it to climb out onto the land. It is particularly fond of grazing on the alga-grown edges of the tidal pools or else hunts for small crabs. It keeps its skin moist by rolling the whole body to and fro in the water at the pool's edge, so that it can breathe through its skin. Frequently it also does a short dive into the water. In the event of danger, it can jump a yard or so to reach the safety of the sea.

I have many times observed in the Indian Ocean how well the mudskipper knows its surroundings. Clearly aware of its objective, it hops through many tidal pools to reach the sea. No detour is too much if it helps to make its flight safer. It gives the first impulse for its leaps by a sort of catapult action of the rear body. Another factor which considerably assists this skilful method of retreat is the possession of highly-developed eyes, which the specialized structure of the cornea makes equally suitable for seeing in water or on land.

These few examples will already have made it clear that survival in the tidal zone is primarily a physiological problem. The body must be protected against drying out under the action of the hot rays of the sun. But adaptations in behaviour are also necessary, as demonstrated by the leaps to safety of *Periophthalmus.* The tidal zone can be colonized either from the sea or from the land. The water creature is faced with the problem of finding the protection of a tidal pool at low tide, whereas the dangers which threaten a land creature occur with the incoming high tide. Two examples will enable us to see clearly how a marine creature and a land creature have each colonized the tidal zone, without abandoning their aquatic or terrestrial mode of life.

How the brittle-star settled in the tidal zone

Around the Indo-Pacific coasts, a person walking along the beach can all too easily overlook the innumerable tiny arms of the browny-black brittle-star *(Ophiocoma scolopendrina)*, which sits along the edges of the tidal pools, making remarkable wobbling and rowing motions with its arms (Plate 20). Why has the brittle-star run the risk of penetrating into the tidal zone? Would it not have been just as satisfactory to stay deep down in the coral reef where the ecological conditions are less difficult?

Every brittle-star of this species has its own home. It occupies a dwelling-hole in clefts in the rock or in flat pits which it does not leave for weeks at a time. In studying these creatures, Prof. Dietrich Magnus observed a facet of behaviour unknown in the echinoderms. This was that they chased away members of their species and neighbours of other species with pushing, sliding movements of their arms. Even when feeding, the brittle-star does not leave its home. It always maintains the central portion of the body or at least the tip of one arm in its refuge. If its grazing-grounds become dry at low tide, it retires into its hole, which always contains enough water to enable it to live through the low-tide periods. These protected places also offer it complete safety when the advancing waves sweep over its home as the tide rises. Brittle-stars are sensitive to movements of the water. Consequently they always remain hidden at high tide and retract themselves as far as possible into their dwelling-holes. They only go foraging during the low-tide period. Here too Magnus discovered a series of astonishing behavioural adaptations.

The brittle-star is not a specialized feeder, for this would offer little advantage in its particular situation. It consumes literally everything that comes its way. Its principal food is green algae, which grow very well here in the tidal zone. In deep tidal pools which never dry out completely, these creatures graze on the meadows of algae in undisturbed water. While doing so they hold tight to the ground with their tiny tube-feet and crop off the growths by bending their arms. *Ophiocoma* has an excellent sense of taste and always tests the quality of what it is about to eat. Once gathered, the food is transported by the arms to the mouth with astonishing skill. But not every brittle-star succeeds in finding a deep, peaceful tidal pool. Most of them have to be content with less ideal spots, the surroundings of which become dry from time to time. In order to survive here, the creatures have developed other feeding techniques. They extend their arms rigidly upwards into the flowing water, the tiny tube-feet on the underside of the arms being braced at a right angle. In this way they form a comb-like filter, which traps the particles carried by the flowing current. During this filtering operation these brittle-stars never entirely leave their holes, always securing their retreat by means of one arm left in the hole to provide an anchorage. Since filtering does not always yield a very high harvest, *Ophiocoma* have also developed yet another method, the application of which is, however, contingent on certain ecological conditions.

At low tide the sun dries out on the surface of the exposed seabed an exceedingly thin layer of plant and animal residues which is floated off on the returning tide. This is a dry food in concentrated form, floating on the surface of the water. At the moment when the water submerges them, the brittle-stars leave their holes, and – moving in the layer of water, which is only a few centimetres thick – come into contact with the layer of dust on the surface. They turn over on their backs, so that their oral side is turned towards the surface. Now they perform swinging movements with the arms, the tips of the arms striking each other in a synchronized movement, so that the film of dust lying between the arms is firmly pressed between them, compacted into a sort of food sausage by means of mucus and passed to the mouth along the arms. The arm movements must be rapid and very well coordinated, since the rapidly-rising tide can wash the film of fine particles away. If we compare the mode of life of this brittle-star with that of other species, we see that it has been a pioneer among the brittle-stars in respect of both its territorial and its homing behaviour. It is astoundingly versatile in feeding habits and higly adapted to the conditions it finds around it. This has presumably made it possible for *Ophiocoma* to make the transition to the tidal zone, where it has access to new sources of nourishment. This is why the brittle-star has left the protected and peaceful hiding-places of the coral reef and ventured onto the land.

The world of the tidal lizard

On certain volcanic islands in the Indo-Pacific there lives a small lizard only 10 centimetres long (4 inches) which bears the name *Cryptoblepharus boutoni.* By day it leaves its refuge on the land and enters the tidal zone. Why should this small, smooth lizard have invaded this zone? Is it a miniature edition of the famous Galapagos marine iguana which has become adapted for life in the tidal zone?

To study the behaviour of these lizards, and in particular their adaptation to tidal zone life, I observed them for a considerable period on a small island off Madagascar.

By night these lizards lodge above the tideline in the fissures of the black basalt rock, issuing by day to go on their peregrinations – following the water line – in the tidal area. In the process they exhibit marked aggression towards members of the same species, which they will not tolerate in their vicinity. Many individuals indeed are without tails, which have been bitten off by other members of their species. In the tidal zone they rummage in every accessible fissure and cavity. They skilfully avoid the tidal pools and even when pursued they will not enter the water. I discovered that these lizards are very fond of hunting the mudskipper *(Periophthalmus).* These fishes frequently sit in shallow water directly below the surface of the tidal pools. The lizards creep very carefully towards their prey and take them by surprise with a single leap. This often causes them to fall involuntarily head-over-heels into the water, revealing in the process that they are exceedingly good swimmers.

To observe the paths followed by the lizards I marked certain individuals with red nail polish. It became clear that these creatures are relatively faithful to one spot and in heavily fissured regions always follow the same path back out of the tidal zone. Moreover, at a fixed time in the afternoon – independent of the state of the tide – they leave the tidal region. One individual I observed exhibited a difference of only ten minutes in the times at which he started his daily return journey.

This timely retreat eliminates the danger of the lizards being overwhelmed by the rising tide during the night. On their way back home they give proof of an astonishing knowledge of their locality. It would be an easy matter for them, if they did not know the terrain, to move onto peninsulas in the tidal zone, which would then become completely surrounded as the tide rose further and finally be submerged altogether. How do the lizards know when there is no danger for them as they enter the tidal zone? Probably they rely on the moisture content of the ground, as they find it when they begin their expedition. They incessantly use their forked tongues to probe the ground before them, and the ground surface has to be relatively dry before they start their movements. If it has rained and the surface of the ground is still moist in spite of the tide being low, they move into the tidal zone only hesitantly.

The lizards also go on extensive journeys to explore the terrain around them, but always find their way back to their starting-point. When I took twenty marked lizards in small cardboard boxes and released them outside their normal area, nineteen of the twenty returned to their base. This was all the more surprising because on the return journey they had to traverse extensive sandy areas. My observations showed that the lizards have adapted themselves to the periodic change in the tides. The homing capacity, combined with precise knowledge of the terrain, enables them to return to land as soon as the return of the tide makes the conditions in the tidal zone unfavourable to them.

A similar behaviour has already come to our notice in the case of the *Planaxis* snails and the brittle-stars. The existence of the homing faculty has been demonstrated in the case of many of the marine invertebrates and fish living in the tidal zone. Clearly this is a faculty which occurs widely in nature and assists creatures to colonize living-spaces with extremes of conditions. The question as to what may have forced the lizards to penetrate into the tidal zone cannot be answered with any certainty. Robert Mertens, a research worker who has specialized in the study of reptiles, is of the opinion that it was excessive pressure from enemies on land which forced the lizard to invade the tidal region, where there are no hostile reptiles. I prefer to believe that, since the lizards inhabit bare volcanic islands, the impulsion to find new sources of food has led them to penetrate into the tidal zone. Even the large Galapagos marine iguana, which has no reptilian enemies on land, lives on volcanic islands the surface of which affords little nourishment. These creatures graze in the tidal zone on the huge meadows of algae, and also return to land at high tide. But these iguanas are already much more highly specialized in their adaptation to life in the tidal zone than is the case with the small *Cryptoblepharus* lizards. The Galapagos iguanas have already reached the point of diving into the sea to seek their food under water. What is more, the high tide does not necessarily force them back to land. Tucked well down and holding tightly on to the rocky surface, the iguana allows the great Pacific breakers to roll over it without effect. As against this, *Cryptoblepharus* is decidedly water-shy. Although it finds its food in the sea, it has remained a purely terrestrial creature.

The sand lagoons and the reef ridge

In the brightly-lit, sandy lagoons, with their sparse growth of coral, we find again certain creatures which also occur along the reef edge. Butterflyfishes (Chaetodontidae) and surgeonfishes (Acanthuridae) gnaw at the coral and feed on the algae which are particularly lush here. A striking number of young fishes are to be found between the coral growths, where they hide for shelter among the – frequently dead – branches. The sandy seabed is combed for food by goatfish *(Pseudupeneus).* They use their whiskers, which contain olfactory organs, to sweep the bottom for particles of food. In addition there are numerous gobies (Gobiidae) and blennies (Blennidae, Plate 31), which are frequently very well camouflaged. Some of them build small caves out of bivalves and dead coral branches.

Soles and plaice lie on the surface of the sand or partly dig themselves in. These fish, which have a remarkably effective camouflage coloration, reveal their presence only by the eyes which project out of the sand. Frequently when I have been swimming across the sand, I have crept very close to them; when they finally did notice me, they shot away, panic-stricken with powerful movements of their fringe of fins. These creatures live in clearly defined regions which they defend against members of the same species.

In the vicinity of dead coral branches on the rocky substrate, colonies of sea-anemones have established themselves, with fishes of the *Amphiprion* and *Premnas* genera and certain types of crab between their tentacles (Plates 64, 94–96). How these fish are able to live among the poisonous tentacles and what adaptations are required to reach this end, we shall see in the chapter entitled 'Partnerships Between Species' (p. 143). The sea-anemones are not so numerous along the edge of the reef, because many of their enemies also live here, enemies which are immune to the venom of the tentacles.

Some sections of the coral are readily colonized by damselfish of the species *Dascyllus aruanus* and *Dascyllus marginatus.* These fish live in groups and are markedly faithful to their chosen spot. In the event of danger they immediately flee to their selected home in the coral and hide among its branches. Even if the coral stock is lifted out of the water, the fish remain in their places.

In the level areas of the coral reef the diver makes his first unpleasant acquaintance with the spines of the diadem sea-urchin, which can be as much as 40 centimetres (15 inches) in length. These sea-urchins congregate in large groups by day, interlocking their spines with one another (Plate 139). Individually, these creatures hide in fissures in the vicinity of the reef ridge. Diadem sea-urchins also frequently inhabit the broad sandy areas which lie at depth at the foot of the reef. Many cardinalfishes (Apogonidae) seek refuge between the spines (Plate 90). They draw back as far as the body of the sea-urchin if you gently move your hand across the forest of spines. A short-spined relative, the *Astropyga* sea-urchin, bears luminescent blue 'eyes' on its dark-violet to brown body. These consist of specialized pigment cells which are highly sensitive to light. Using its spines the *Astropyga* can creep very rapidly across the sea-bed. Sometimes these creatures have approached the black housing of my underwater camera, mistaking it for a member of their own species. The congregation of groups of these creatures is in fact initiated by a dark-orientation phenomenon (skototaxis).

Of other sea-urchins, for example the poisonous *Tripneustes* or *Echinothrix,* frequently the only thing visible is a few spines projecting between patches of algae, bivalve shells and pieces of dead coral. The sea-urchins can in fact hide themselves completely. It used to be believed that this was their manner of camouflaging themselves or protecting themselves from the light, but the phenomenon is nowadays explained by reflex movements of the tube-feet and spines. The piling-up of small objects over the urchin is triggered off by the same reflexes as those which cause it to move forward. A particularly specialized species of sea-urchin, the sand-dollar, with spines only a millimetre or so in length, lives buried in the sand; it glides over the sea-bed like a tractor, picking up particles of food from the sand and carrying them, by cilia on the body surface, to the mouth.

The transition zone from the sandy lagoons to the reef ridge is often formed by a zone of algae, which are also colonized by forms of organisms specially adapted to this zone. Many species of lipfish live among these algae, their coloration being very close to that of their surroundings. One master of camouflage is the slender cigar wrasse *(Cheilio inermis)*, which can assume various colour patterns to suit the surroundings of the moment. Among the algae, these fish are grass-green. Often the cigar wrasse 'rides' close to goatfishes (Plate 89), letting them guide it over long distances. In this case the wrasse bears a clear yellow livery. A diver will rarely succeed in identifying in the forest of algae the venomous scorpionfish (Scorpaenidae), which are rendered almost invisible by their outstanding camouflage. These fish wait patiently for an opportunity to seize their prey, sucking it with lightning rapidity into their enormous gaping jaws.

The old reef ridge which has died off on the coast side is covered by a thin layer of fibrous algae, on which the butterflyfishes (Plates 48, 50–54, 59) and surgeonfishes (Plates 27, 66) love to graze. These fish have teeth specialized for this purpose, enabling them to graze freely on the algal growths. At high tide the submerged ridge of the reef is regularly visited, in search for food, by fish which otherwise live only along the reef edge.

In the ramifications of the fissures and holes are the small sea-urchins *(Echinothrix calamaris)* which – like most sea-urchins – pick over the surface of the rock during the night. Deep in the crevices of the reef rock we also find the magnificently coloured sea-urchin *Heterocentrotus*. This species possesses particularly thick spines, which have sharp edges at their outer ends. Over a period of time this urchin bores a living-hole in the coral rock and frequently cannot get out again; it is also found in fissures in the steep seaward precipice of the reef.

One inhabitant of the reef ridge is *Tridacna*, the giant clam, which embeds itself firmly into the coral rock (Plate 29), and which on the Great Barrier Reef of Australia attains a shell length of 1½ metres (5 feet) and a weight of 200 kilograms (440 pounds). Its shining coloured body lies clearly revealed between the two valves of the shell, shimmering in all the colours of the rainbow. A striking feature is the large siphon aperture in the body tissue, through which the water sucked in is expelled in a jet. As the shadow of a diver falls across the giant clam, the two valves of the shell snap shut. Its power of closure is so great that it is possible to open the valve only by means of a crowbar. The power of a *Tridacna* is ample for it to hold a man if he were to be negligent enough to put a foot between the two valves. However, accidents of this kind only occur in films. The rapidity of closure is so great that it would be very difficult to put one's foot between the two valves of the shell.

Many fish utilize the fissures of the reef ridge as refuges. It is here that we frequently meet the Picasso fish, *Rhinecanthus aculeatus.* By means of specialized spiny dorsal-fin rays this fish can clamp itself into the fissures so firmly that it cannot be extricated even by the use of considerable force. In the South Seas, it has become a somewhat macabre sport for many tourists to hold races with such fish. A cork is impaled on the clamping spine, so that the fish can no longer dive, with the result that it paddles along helplessly close below the surface, and its 'racehorse' progress can be clearly observed from above. The man with the fastest Picasso fish gets a prize.

In the clefts in the reef ridge, hidden deep in the rock, many fish sleep by day which are purely nocturnal in their feeding habits. Often, for instance, we encounter the surgeonfishes *Acanthurus sohal* (Plate 27) or *Acanthurus lineatus,* so called because of the razor-sharp bony keel which they carry at the root of their tail and can erect like the blade of a penknife. Many fishermen are said to have been injured by these fish, which constitute typical specific forms of adaptation to conditions in this zone of the reef. They glide elegantly through the numerous crevices in the reef or skilfully evade the foaming surf. The flat body shape is doubtless an adaptation to enable them to manoeuvre rapidly in the unpredictable confusion of fissures, crevices and coral branches.

The edge of the reef and the reef precipice

Along the edge of the reef the full variety of marine life unfolds. Here we encounter the most varied kinds of coral. The huge fans of the stinging or fire coral *Millepora* (Plate 34) – which the diver should do everything he can to avoid touching – resemble true corals closely enough to be mistaken for them. The name fire coral has wrongly become attached to this type, wrongly because they do not come under the category of corals (Anthozoa), but are stock-forming hydrozoa. The nettlelike stinging action of the polyps is so strong that the skin turns red and swells up, and the sting is frequently accompanied by high fever. The fans of fire coral generally stand at right angles to the direction of the water current.

Hundreds of small damselfishes belonging to the genera *Chromis* and *Pomacentrus* swim in the open water above the coral growth and snap up the small particles carried by the current. Many species of fish live here in very close contiguity to each other. Primarily to avoid confusion of species at mating time, nature has provided each species with its own unmistakable coloration, of which I shall have more to say later.

The red *Anthias* perch form dense schools along the edge of the reef. These fish constitute for me an important indication of the approach of danger. If they are swimming playfully about in the open water, I can work on without any misgivings.

If, however, the school suddenly shrinks together and takes flight to the edge of the reef itself, this is always an alarm signal – a shark or a large grouper is swimming by. It is easy to distinguish between two groups among these fish; one group is completely red in colour, while the others bear on their more rust-brown sides a large golden-yellow shimmering patch. These are males and females, swimming in separate shoals. During the course of their development, the fishes change sex. At the beginning of their life they seem to prefer the role of the female, and it is only later that they turn into males. Many lipfishes have the same capacity for a change of sex. The small *Serranellus* perch from Florida goes so far as to be male and female at the same time. It can occur that at spawning the fish is first of all a male, but directly after this assumes the function of the female by expelling the eggs.

The surface of the coral is nibbled away by numerous parrotfishes (Plate 9). When they are busy chewing at the coral with their hard beaks, the grating noise can be heard a long way away. They reject the indigestible calcareous constituents and the lime falls to the sea-bed in finely-ground form. In this way certain species of fish play their part in the production of sediment. The puffers and triggerfishes also gnaw at the coral. Specialized in most versatile ways are the very brightly coloured angelfishes, emperor-fishes and butterflyfishes (Plates 48, 50–54, 58, 59). The first of them swim about in pairs and clearly have mated for life. Some species have long tweezer-like beaks with which they can probe into deep fissures in their search for small crabs, worms and snails; one particularly striking species is the long-nosed butterflyfish *(Chelmon rostratus)* with a huge eye-spot on the dorsal fin, another is the shining yellow *Forci-piger longirostris.* With their long thin snouts they can suck in small prey as though with a pipette. All these fish are specialized in different ways. They have some special preference for a given kind of food or for a particular living-site, or are specialized in behaviour.

At first sight the world of the reef strikes the diver as being peaceful and harmonious, and only seldom does one observe fighting. But appearances are deceptive. All these creatures are strongly territorial and respect an invisible boundary between themselves and their neighbours. The atmosphere is charged with tension, none of which is evident to the diver. It is only when a fish forces its way out of its district, so disturbing the equilibrium of forces, that the unrest breaks out visibly. This is when fighting really does occur, since it is essential to drive the invader away from the territories it has invaded (Plate 49).

Whereas members of the same species are rivals, who must be kept out of one's own territory, each individual has also to protect itself from predators which seek its life. Various species of grouper *(Cephalopholis* and *Epinephelus)* and the reef perch *Aethalo-perca* hide in various cracks in the rock, coming out every now and then to seize their prey among a shoal of small fish (Plates 135–137). A great many predators who attack by surprise are, in addition, exceedingly well camouflaged and can approach their prey unnoticed. Notorious among these are the scorpionfishes, to which reference has already been made and which can live in almost all zones of the reef. The most dangerous of these, the stonefish (Plate 73), can hide itself so perfectly by means of its camouflage that it cannot be distinguished from its surroundings at all. Lying in wait on the sea-bed it can, if threatened, use the spikes on its dorsal fins to sting, injecting into the fish a deadly poison, which has even proved fatal to human beings.

In the Red Sea there lives a green anglerfish *(Antennarius)* which not only uses camouflage to hide itself from its prey, but also lures it within range. A deceptively worm-like process growing from the forehead serves as a lure: the victims swim towards this presumed titbit and are immediately drawn into the anglerfish's huge throat.

Along the reef precipice the long-spined turkeyfish *(Pterois volitans)* takes up its place; this fish has the pectoral fins divided after the fashion of wings (Plates 109–110). The spines on the dorsal fins ripple in a wavy movement and can be brought into play very fast to sting in the event of danger, injecting a powerful poison which causes great pain. By day, the turkeyfish remain hidden in cracks in the tock, their fins folded away. Only as evening falls do they become active and go out hunting. They stalk their prey slowly, drive it into a dead end with their wing-like pectorals and swallow it in a trice.

The muraena or moray eel (Plate 65) hides in recesses in the rock, and it too hunts predominantly at night. Exceedingly well developed scent organs enable it to detect its prey. But some fish do manage to escape them by surrounding themselves at night in a mucilaginous sleeping-bag, which is produced by a gland situated behind the gill covers. This envelope of mucilage protects the fish against the olfactory apparatus of the moray eels.

The many small gregarious fish which inhabit the edge of the reef and the precipice wall are not only pursued by the predators who have made their homes there, but also by many other predators from the open water. Sharks, barracudas and spiny jack *(Caranx)* come to the reef at fixed times to carry out their unremitting hunt (Plates 147, 148, 150).

If we swim further down the reef precipice, we find the large eyes of soldierfishes and cardinalfishes staring at us from dark crevices (Plate 38). Some of these even utter sounds, which divers under water can hear very clearly. In deeper caverns grow magnificently coloured red algae, sponges, bryozoans, gorgonians (sea-fans) or even the beautiful Alcyonarian corals, which are predominantly reddish or yellow in tone (Plates 6, 7, 30). Numerous brittle-stars cover the walls of the caves. One striking sight is the brilliantly-coloured thorny oyster *(Spondylus)*, which closes its valves as soon as a shadow passes over it.

The fish which live in the caves always turn their bellies towards the wall of the cavern. Normally fish orientate themselves with their backs to the light. However, if light is absent, as is the case here in these undersea caves, they take up a position governed by the structure of their surroundings (Plate 40).

These caves also serve as refuges for the large spiny lobsters *(Panulirus)*; these creatures are easily recognized by their long feelers which project out of their holes (Plate 41). During the American underwater-living experiment off the Virgin Islands in the Caribbean in 1970, in which teams of marine biologists lived and worked for 2–3 weeks at a time in the *Tektite* sea-bottom laboratory, spiny lobsters were equipped with small radio transmitters, to enable them to be observed over long periods. Shortly after sundown, they leave their holes, and go off foraging on the sandy areas at the foot of the reef. One lobster covered 240 metres (nearly 800 feet) in the space of 24 hours. On one occasion a group of spiny lobsters disappeared for three days, after a school of seven large nurse sharks had arrived in the neighbourhood; it is uncertain whether the departure of the lobsters from the area was caused by the sharks.

It is not only the large cavities and crevices in the reef precipice which are inhabited, but even the very small interstices in the coral offer dwelling-places for many creatures. Between the branches of the acropora live porcelain crabs *(Porcellana)*, or sculpins *(Caracanthus)* which attach themselves tightly between the coral branches by means of the spines on their gill covers. Gobies can also specialize themselves to live among the coral branches. Other fish again, like the small cigar wrasse *(Chromis caeruleus)* and the damselfish *(Dascyllus marginatus)*, use the interstices as hiding-places in the event of danger. Frequently some coral branches are thickened as though by galls. If we look closely at these, we can see either a thin crack or a ring of folds dividing this enlarged growth into two parts. In the cavity within the gall there is a female of the *Hapalocarcinus* or coral-gall crab. The crab can no longer leave its dwelling-place, because the coral has grown around it. Food flows in through the holes or a gap. But how can the crab possibly multiply, if the female is walled-in for the rest of her life? To this end nature has devised a trick – the male is very much smaller in size, so that he can easily creep through the holes to reach the female and mate with her.

It is not only on the surface or even in the cavities between the coral that animals have settled; the interior of the calcareous skeleton is also thickly colonized by algae, sponges, boring snails, bivalves and worms equipped with specialized boring and digging tools. Some of them even dissolve the chalk by chemical means. In addition to the various zones of the reef, the coral itself is also inhabited by various highly specialized species.

The coral debris, sand and mud zones

Dead coral debris accumulates at the foot of the reef precipice. This debris zone is a favourite place for the fish that inhabit the steep reef wall, which come here to crop the algae. Scorpionfish inhabit the many holes in the rock in this area too. A squat, broad-backed reptilian-looking fish which completely melts into its surroundings by reason of its body shape and coloration, it has a gaping mouth which betrays the fact that it is a rapacious predator, lying in wait here for its prey, which it overwhelms in one lightning jump.

We cross this zone of debris to swim over wide open areas of sand which in places gives way to mud. Like the creatures of the beach zone, the inhabitants of these bare regions can easily be seen by their enemies. Here again the fish and other creatures are thereby forced to adopt particular adaptations. The best protection against enemies is to burrow under the sand. This, for instance, is the technique adopted by many sea-cucumbers, relatives of the starfish and sea-urchins, who by day dig themselves in below the sand and come out only at night. Some sea-anemones have also become nocturnal. As soon as they are touched, they retract extremely rapidly into the ground. One creature only recently discovered and investigated builds on the sea-bottom conical sand-heaps like those of the horseman crab on the beach (Plate 45). Frequently these little sand-hills seem to smoke like a volcano, and a fountain of sand appears briefly at the apex when this happens. Next to each sand-hill is a funnel, which terminates below the sea-bottom in a tube. The sand-hills and the funnels are connected by a system of tunnels, the funnel apparently serving as a trap. Sand containing particles of food falls into it, and is then eaten, the indigestible residues later being discharged from the apex of the cones. I was able to examine the course of the underground tunnels very precisely. The tubular tunnel is about 3 to 5 centimetres (1 to 2 inches) in diameter, and the walls are firmly cemented with mud. However, despite my energetic digging, I have never been able to catch a glimpse of the creature that made these holes; I once dug a hole a couple of yards deep in the sea-bottom. Cousteau caught a crab which he supposed to be the volcano-builder. It has however not been possible to confirm this hypothesis as yet.

Many diurnal creatures build burrows or tubes in which they take refuge on the approach of an enemy. The jawfishes *Opistognathus* and *Gnathypops* make strong burrows of bivalve shells and pebbles, skilfully lining the burrow with the chosen building material. The shells and stones are piled singly one on top of another, so that the wall of the burrow will not collapse.

The sand-eels *(Trichonotus)* live together in small groups and dive head-first into the sea-bottom when danger threatens. They bore through the sand and pop their heads out elsewhere. Another inhabitant of the sandy bottom, the slender *Gunellichthys monostigma* (Plate 47), builds a tube only 5 to 10 millimetres (0.2 to 0.4 inch) in diameter. This fish always remains close to its tube and fishes for the plankton carried past by the water currents. At the slightest sign of danger it flees into its hole. On several occasions I have carried out the experiment of blocking the hole with sand. Immediately the fish lost all sense of direction and fled in panic. In so doing it encroached on the territories of other members of its species and was attacked by each of them.

The well-known garden eels (Heterocongridae) also retire into tubular burrows of their own manufacture in the event of danger. With extreme rapidity they wriggle backwards into their holes. These garden eels live in large colonies, which have the appearance of an asparagus plot. As many as five thousand individual eels can be present in a single colony. The tubes are dug by means of short jabs with the tip of the tail, which has become adapted to the sand-haunting habits of its owner and takes the form of a hard, pointed awl. In addition to this, there is a mucus gland situated at the end of the tail, producing a secretion which serves to cement the interior walls of the burrow. Garden eels are a real dainty for predators, but they find them very difficult to catch. The camouflaged lizardfish *Synodus,* to take one example, lies in ambush near a tube and waits for the eel to come forth. But this predator represents a danger only to small eels. In Madagascar I have observed the jackfishes pouncing on eel colonies like diving fighter-planes. Meanwhile, several snappers attempted to attack from the flanks. However, I have never yet observed such attacks achieving any success.

The water above the reef and the open sea

In the clear water above the reef we once more find closely-knit associations. Many reef fishes pass the juvenile stage of their life drifting as small free-floating larvae among the plankton, and only subsequently return to the reef. Others swim out into the open water only for feeding. Fish of the genus *Caesio* swim around in huge shoals and feed on plankton just off the reef. In the evening they return to the reef, sleeping on the sea-bottom near the coral. The dark-violet triggerfish *Odonus niger,* too, form small groups some little way away from the reef. Each fish has a particular hiding-place in the reef, into which it slips when danger threatens, and in addition, sleeps in this hole at night. At spawning-time the triggerfishes are particularly aggressive and even attack passing divers. They have powerful jaws, and their orange-coloured teeth can punch deep holes into the diver's suit and, in addition, cause painful injuries.

However, many other fish spend the whole of their lives in the water above the reef. Close below the surface of the sea hunt the garfish *(Strongylura)* and halfbeak *(Hemiramphus),* preying on the small fish which abound in shoals. These predators frequently hunt in packs and carry out regular battues. Even the large predatory jackfishes swim in twos or threes, and fall upon their prey in sudden lightning-like surprise attacks. The fastmoving bonitos also hunt in a similar manner. Finally, there are some species of shark which inhabit the water above the reef. The dogfish *(Triaenodon obesus)* generally swims near the sea-bottom, but seeks its prey along the steep reef wall. The blacktip reef shark *(Carcharinus melanopterus)* and the grey shark *(Carcharinus menisorrah),* on the other hand, generally patrol the open waters in the vicinity of the reef precipice. Open-sea sharks come into the reef from time to time in search of prey. In fact the tiger shark will even swim into shallow bays where the depth of water is only a few feet. The inhabitants of the reef are thrown into a state of considerable alarm by the approach of a shark and immediately withdraw into their refuges.

A good way off from the reef the pelagic associations, the life associations of the open sea, begin; these are the creatures which are independent of the sea-bottom and live their whole lives out in the endless blue expanses of the ocean. To prevent them from sinking into the depths, many have developed feather-like body processes which can keep them floating freely. Others store in their bodies drops of an oil which is lighter than water.

The strange medusae are almost 99 per cent water, and maintain themselves in a free-floating state by pumping, pulsing movements of their 'umbrella'. Around the edge of the umbrella they have simple orientation organs. Since they do not have to perform any difficult movements, their sense organs are minimal. A research worker has said that the medusae really only perceive the stroke of their own bell. Some medusae possess gas-filled floats, by which they maintain themselves on the surface, like the Portuguese man-of-war, *Physalia.* These creatures trap plankton with their exceedingly long tentacles which can extend to as much as 50 metres (160 feet), plankton being the staple diet of most of the creatures that inhabit the open sea; even the largest fish alive today, the whale shark (15–20 metres, or 50–65 feet, in length) or the manta ray – which can reach a 'wing-span' of up to 8 metres (25 feet) – live on plankton. The smaller fishes generally collect in schools, since the individual is safer in a group of this type (see Chapter 7 on schooling behaviour).

The large predators of the high seas, such as the tiger shark, blue shark and white-tip shark, have greater difficulties in obtaining their food. They have to maintain a continuous hunting patrol in order to assuage their hunger. The high seas they inhabit do not offer them such abundance of food as is available to their reef-hunting relatives who live, comparatively, in abundance.

Coral reefs in danger

The corals, which are very finicky in their choice of site, are very dependent upon the surrounding conditions, where many dangers constantly beset them. They are exposed without defence to the destructive power of the surf. Hurricanes pass over many coral islands and destroy large sections of the fauna. Fresh water in the vicinity of the coast can also cause considerable damage.

Along the reef itself there are numerous creatures which destroy the coral. We have already met those busy coral-eaters the parrotfishes and triggerfishes. Snails too are among the corals' foes. The gastropod *Jenneria* consumes 1 gram of coral material per day. Up to ten thousand individuals of this genus can be present over an area of 1 hectare (2½ acres). The small crab *Trizopagurus* eats only 10 milligrams a day, but there can be as many as 250,000 individuals per hectare, according to American scientists.

One particularly dangerous coral-eater is the multi-armed, spiny crown of thorns *(Acanthaster plancii)* (Plates 12–14). This can reach a spread of 60 centimetres (2 feet) and in a single month can eat about 1 square metre (10 square feet) of coral polyp. In recent years this giant starfish has multiplied so greatly in the Pacific region as to become a very serious threat to coral reefs. The damage which has occurred hitherto is alarming and it is still not possible to predict the ultimate consequences. In the summer of 1969, the Pacific Reef Starfish Expedition – comprising eighty marine biologists – was therefore despatched to Micronesia to study the behaviour of the crown of thorns and the degree of reef damage. The scientists established that on the islands of Guam, Ponape, Rota and Saipan thousands of crown of thorns starfish had totally destroyed the reef over stretches often up to a kilometre in length. As much as 200 kilometres (125 miles) of the Barrier Reef are said to have been destroyed already. Counts were carried out in Australia by marine biologists to estimate the density of population of these creatures. One Australian biologist swam for 1 hour 40 minutes through a reef on Green Island, and counted more than 5,750 individuals. This starfish plague assumed such proportions that the Government of Queensland set up a special committee and provided funds for research into the life pattern of *Acanthaster.* Research workers are also studying the question of whether reefs which have once been destroyed can regenerate themselves and whether a new coral growth will occur. The fact is that reefs which have been destroyed rapidly become overgrown with green algae, which can prevent any resettlement by coral polyps. Considerable harm has been done to the fishing industry as a result of the depletion of the fish population in the vicinity of dead reefs. The fear has also been expressed that small atolls, to which the coral reefs serve as natural breakwaters, will now be more easily eroded by wave action.

What is the reason for this mass growth of the starfish? Several hypotheses have been put forward. The triton snail *(Charonia tritonis)* is a natural enemy of *Acanthaster.* Tourists have collected this beautiful snail in large numbers from many places around Australia and it has therefore been thought that this destruction of its natural enemy has allowed *Acanthaster* to multiply rapidly. However, one piece of evidence which goes against this is that there are many coral reefs where tritons have also been collected, without any growth of the starfish population. Some scientists hold the view that underwater explosions and other man-made disturbances in the sea-bottom life killed off many plankton-filtering creatures which also trap the tiny free-floating larvae of *Acanthaster.* They suggest that the surface of the sea-bottom is now free of creatures which would feed upon the larvae, so that more *Acanthaster* larvae reach the breeding stage. Another possible explanation is a rise of temperature in the Western Pacific Ocean shortly before the first massive rise in starfish population. Finally, a mutation theory has even been advanced, according to which the genetic apparatus of the starfish has undergone such major modifications that it is now better adapted to its surroundings. It is however extremely improbable that such 'revolutionary' genetic changes in the heredity pattern should have occurred simultaneously in many parts of the Pacific.

According to recent reports, mass population increases of the crown of thorns have been observed on previous occasions. This indicates the possibility that certain combinations of ecological conditions (enemies, food supply, parasites, physico-chemical factors etc.), which control the population density of the starfish, can periodically produce exceptionally good breeding years. If the present massive population growth is the result of such periodic influences, it may be hoped that the population density of the starfish will in the future revert to a normal level.

The present control measures are simple, but not sufficiently effective. Starfish are collected by divers or killed underwater by injections of formalin. When I visited the island of Okinawa recently, I observed damaged areas in the reef. Swimmers who dive for sport off Okinawa have hitherto collected more than two thousand starfish. Only research into the manner of life of the crown of thorns will make it possible for us to develop biological methods of control and to deploy them appropriately.

Normally, natural selection ensures the ecological equilibrium in the reef, so that the enemies of the reef-dwellers do not predominate. It is only when this equilibrium is disturbed that the enemies suddenly become a serious source of danger. American scientists believe that a reef which has once been destroyed has little chance of regeneration, since after passing a certain 'critical phase' the destruction of the reef proceeds at an increasing rate, because the surface of attack presented to the enemies is growing larger all the time. In the last few years, Man has also revealed himself as a dangerous destroyer of the ecological equilibrium of the coral reef. Underwater tourists armed with harpoons have not only decimated the fish population of many reefs, but have also damaged the reefs themselves by senseless plundering of the coral fauna. Everybody wants to take home a souvenir from the depths of the tropical ocean. A coral branch broken off from the reef soon begins to smell unpleasantly once it has been brought out into the air and it is then very quickly discarded. In Hurghada on the Red Sea, the building of a hotel gave large numbers of tourists the opportunity of diving in the coral reef. I have personally been able to observe how, in the course of a few years, the once flourishing reefs have been destroyed. Unhappily, many states only protect their coral reefs after considerable damage has already been done, as was the case in Hurghada. A coral reef is an ecological area which is highly sensitive to disturbance, and damage once caused can hardly be remedied – not even by belated nature-protection measures.

We know that the oceans have today become the world's dustbin. Cousteau has prophesied that our grandchildren will never see a coral reef, if we do not fundamentally alter our attitude to the sea. It is imperative that we should understand once and for all that the oceans – and this includes the coral reefs – are a part of the world in which we live and that our conquest of this world imposes upon us the responsibility to protect it.

48 *(Opposite)* Colours serve as a means of maintaining the species. The butterflyfish *(Chaetodon austriacus)* from the Red Sea generally swim about in pairs. Their striking body coloration and the fact that they go about in pairs helps to maintain the species.

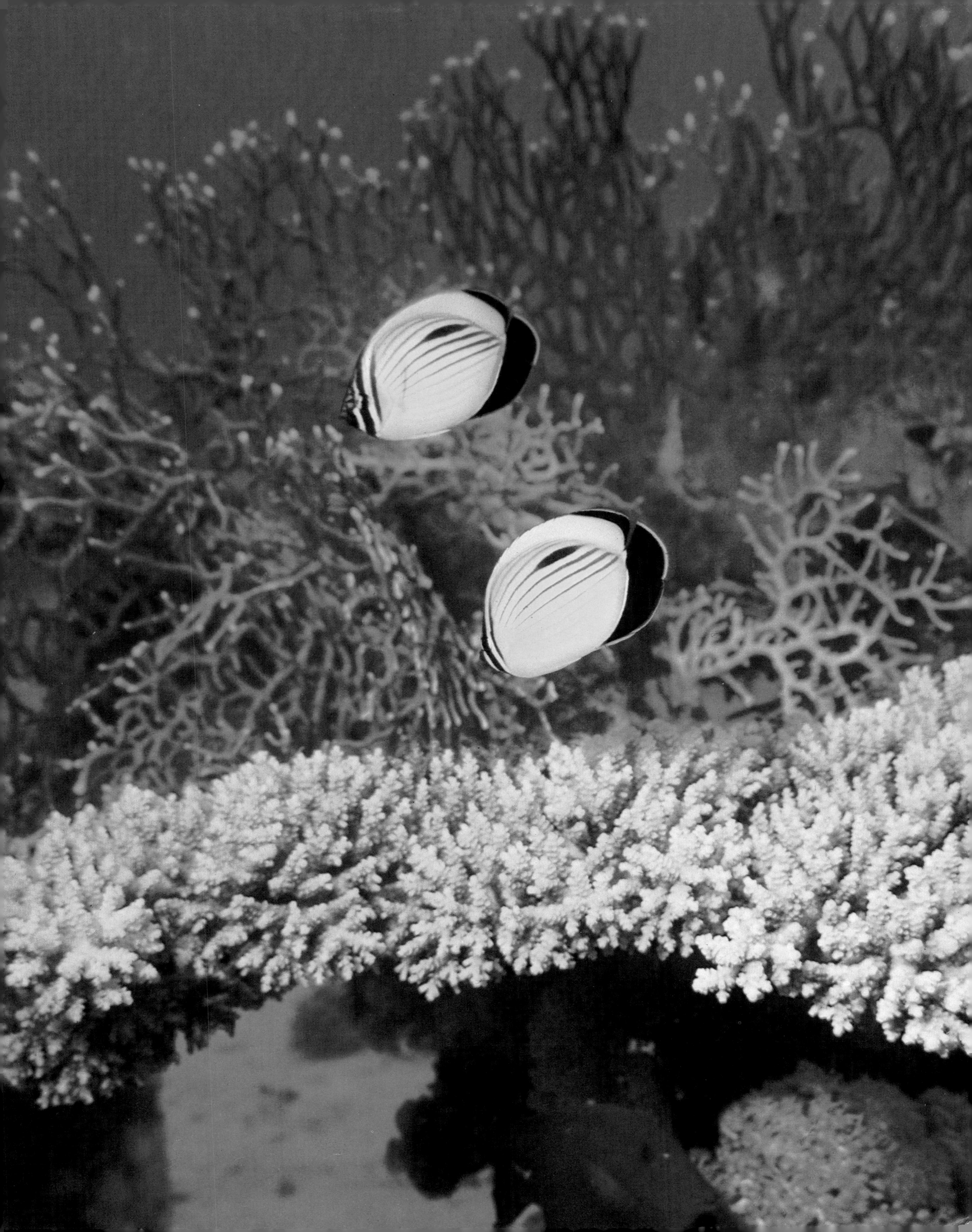

◁49 The labrid *(Lienardella fasciatus)* in the frontal threat position. Most of the bottom-dwelling reef fish stay in one spot and defend it as their own territory. Around the boundaries of these territories 'threat duels' occur which follow a fixed ritual pattern.

50

51

52

50–53 The young (50, 52) of many reef fish have different colours from the adults (51, 53). This striking and eloquent colour pattern is the stimulus that sets off the intraspecific combat behaviour, which produces territorial defence reactions. The young fish would not be able to defend themselves against the attacks of the adults unless they were 'masked'. Top: French angelfish *(Pomacanthus paru)*. Below: Imperial angelfish *(Pomacanthus imperator)*.

53

◁54 The butterflyfishes – in this case *Chaetodon chrysurus* – are represented by a large number of species present along the reef. The various species are specialized in different ways. Their striking, vivid colours are the identification signals, directed at their rivals but also aimed at sexual partners ready for mating. The more distinct and conspicuous the coloration and marking of the fish's body, the less likelihood there is of a mistake in the choice of a partner.

55

56

57

58

55–59 The reef fishes have adapted themselves to their chosen area in the most varied manners possible, and they are, in addition, specialized in different ways. Striking signal colorations enable the various non-competing specialists to recognize each other. Whereas the signal colours trigger off aggression between similarly-specialized members of the same species – so contributing to a proper distribution of the individuals over the living-space available to them – these signs also make it possible for different specialists to live in peaceful coexistence. Different forms of specialization therefore favour the adoption of different coloration. In addition, the young and adult members of the same species are specialized in different ways, and they too have different colour patterns (compare Plates 50–53).

55 *Gaterin orientalis* (sweetlips). 56 *Gramma loreto*. 57 *Halichoeres centriquadrus*. 58 *Pygloplites diacanthus* (peacock emperorfish). 59 *Chaetodon semilarvatus* (lemonfish).

59

60, 61 Many reef fish change colour at night. These two photographs show a *Caesio* – while sleeping (60) and by day (61). Presumably the night coloration serves as a camouflage against nocturnal predators.

62 A rare photograph of the coral reef at night. The luminescent fish belonging to the genus *Photoblepharon* bear a light-organ underneath the eye. By means of an eyelid-like 'shutter' of pigment they can interrupt the light emitted by these luminescent organs – in other words, they emit flashing light signals! Two of the fish have closed this pigment shutter. This photograph was taken in shallow water in the Gulf of Aqaba on the coast of the Sinai Peninsula. Nothing is known at present about the significance of these signals.

63 *(Opposite)* Fish are able to signal their mood to other fish of their species by colour changes, as is the case here with the scad *(Priacanthus hamrur)*. This change in coloration is produced by special pigment cells in the epidermis.

60

61

62

65

66

◁64 The striking coloration of the anemonefish *(Amphiprion bicinctus)* is a warning pattern, whereas the large white patches on the young of this damselfish *(Dascyllus trimaculatus)* presumably function as a social recognition signal.

65 The moray eel *(Lycodontis javanicus)* exhibits its beautifully-coloured mouth cavity as a warning signal to an attacking predator. The sudden unveiling of striking patterns serves to frighten off many creatures.

66 Many shoal-forming fish – in this case the surgeonfish *(Acanthurus monroviae)* – wear 'tail-lights' by which members of the same species can recognize them and which stimulate strays to rejoin their group.

67 Communication signals. The neon goby *(Elacatinus oceanops),* visible as a small blue streak in the jaws of the jewfisch *(Epinephelus itajara),* ventures into the 'lion's mouth', searching for parasites. The grouper invites the goby to perform its cleaning function by opening its mouth wide.

◁ 68 The cleaner shrimp *(Stenopus)* – itself hidden – agitates its easily-visible white antennae as an invitation to fish who wish to be cleaned.

69 The rock cod *(Aethaloperca rogaa)* opens its jaws wide to enable the tiny cleaner shrimps *(Leandrites cyrtorhynchus)* to climb inside. Subsequently, the rock cod will indicate by carefully closing its mouth that it is time for the shrimps to leave.

70 *(Overleaf)* The signals given by the cleaners attract fish that wish to be cleaned, and they then assume the 'ready-for-cleaning' position. A wolf in sheep's clothing, the young sabre-toothed blenny (*Runula* sp.), imitates the coloration of young cleaner fish, to enable it to approach its 'client', from whose skin it quickly takes a bite.

2 The language of signals

When I made my first dive to a coral reef more than ten years ago – it was at Hurghada on the Red Sea – there came into my mind as I viewed this abundance of living creatures sweeping by with movements of their fins or tails, floating or creeping past me, the question of why there should be this surprising wealth of beauty. If I am asked today to indicate the purpose of this magnificent range of colours in many corals and sponges, shimmering rainbow colours and exquisite body patterns on bivalves or snails frequently only a few millimetres long, I have to confess that even after ten years of work among the coral reefs I still do not know the answer.

In spite of this I have devoted my attention – encouraged by my mentor Konrad Lorenz – to the multicoloured patterns and designs of some of the reef fish that immediately strike the eye of every diver. There are high-backed frying-pans of imperial yellow darting vigorously between the coral branches, shining blue shapes swimming in shoals above the reef, tiny scintillating jewel-like bodies flitting to and fro above the sea-bottom. Konrad Lorenz, who described these eye-catching fish as having the colours of publicity posters, put his observations into words as follows: 'The garish poster-coloured fish all stay in one place. It is only such fish that I have seen defending their territories. Their raging belligerence is directed only at their fellows, and I have never seen fish of two different species attack each other, however aggressive the two types might be.'

These bright poster-like colours are intraspecific signals, which trigger violent territorial defence reactions between members of the same species. The colours serve to ensure the maintenance of the species and have come about in the struggle for existence. To make this comprehensible, I must now go into some details of the manner of life of these brightly-coloured denizens of the reef.

Specialized feeders

The coral reef exhibits optimum living conditions, making it possible for many fish to find suitable places – so-called ecological niches – there. Some peck away at the soft parts of the coral polyps, others, like the parrotfishes, crunch bivalve shells or nibble at coral branches. The well-known cleaner wrasse *(Labroides dimidiatus)* and the cleaner or neon goby *(Elacatinus)* have specialized in collecting the 'parasites' from other fish. One angelfish from the Caribbean eats sponges containing pointed skeletal elements which are quite unacceptable to other fish. Some specialists, like the large blue trigger-fish *(Balistes fuscus),* even crunch up the bundle of spines of the poisonous diadem sea-urchin. One species of lipfish spends part of its time between the stinging nettle-like tentacles of sea-anemones, from the surface of which it picks off the particles of plankton which have become trapped. One goby *(Cottogobius)* steals the food laboriously gathered by its host, a type of whip coral. All these creatures have adapted themselves in body form and in behaviour to their specialized method of feeding. This specialization, developed in the evolution of the species in question, opens up to them the utilization of sources of food which it is difficult for others to draw upon. Any creature which eats the same food as one of these specialists becomes a threat to it. It is thus not only the predators which menace the life of an individual, but even those of his own kind. The struggle for life unrolls itself in a no less dramatic way between closely related but competing species. If the random interaction of mutation and selection suddenly confers upon a species some new feature, then even another creature which has efficiently adapted itself can be thrust aside.

It is a common occurrence in nature that competing species should become distributed as widely as possible over the space available to them, to ensure maximum utilization of the sources of food and to moderate the acute intraspecific competition. This intraspecific aggression thrusts the individuals of a given species apart, so acting as a spacing mechanism and favouring the wider distribution of the species. Lorenz thus considers the brilliant colours of the 'location-tied' reef fishes (i.e. fishes having a restricted or home range) as being signals which ensure that members of the same species recognize each other and which trigger off attack reactions in the occupants of given territories on sighting a member of the same species. Lorenz observed in the aquarium that the brilliantly coloured fish were always unequivocally more aggressive than the less brightly coloured kinds.

In my observations along the coral reef I have been able to establish that even fish which are not strikingly coloured but are territorial by nature fight viciously as soon as one of their own species approaches. Conversely, there are brightly coloured fish which behave in a thoroughly peaceable manner. The conclusion is therefore that the fact of possessing bright 'poster-like' colours is not necessarily linked with a high degree of aggressiveness. Very bright body patterns, however, are strikingly frequent in the case of families of fish which contain many species – as is, for instance, the case with the butterflyfishes (Chaetodontidae). In coral reefs all round the world these high-backed, flattened fishes occur (they are also known as brushteeth, because they pluck out from the coral surface micro-organisms, polyps and algae, using their brush-like pointed teeth).

I have already mentioned the long-nosed fishes *(Chelmon* and *Forcipiger)* as typical representatives of the brilliantly coloured kinds, recognizable at considerable distances. Their beak-like snouts enable them to suck their tiny prey out of deep hiding-places, which would be inaccessible to other related butterflyfishes.

Having observed these fish in our aquaria at Seewiesen, Germany, I was thrilled at the thought of observing butterflyfish at liberty in the coral reef. How had they become specialized there? What pressure of natural selection had brought about their various patterns and bright combinations of colour?

Observations on butterflyfishes

When I began my investigations along the reef, I marked individual fish with small plastic markers, so as to be able to follow their progress under water and to observe them. In the first few days it became very clear that the butterflyfishes are in fact very faithful to their chosen locality and that they are the most vigorous in defending their territory. At Eilat on the Red Sea I had marked three neighbouring pairs of butterfly-fish *(Heniochus intermedius),* which I have since found again in the very same place over a period of two years. Pair No. 1 of the large left-hand territory regularly comes around evening time to a fallen Acropora coral and waits at this point for its neighbours – Pair No. 2. These two pairs threaten each other and then after a short while slip away again. None of the individuals crosses the invisible territorial boundaries.

The territories occupied by the marked fish vary in size. A single butterflyfish of the species *Chaetodon lineolatus* had taken possession of a section of reef 500 metres (550 yards) long. I knew very precisely his patrolling movements within his territorial district.

During these observations it struck me that butterflyfish take a snack, so to speak, at every corner and apparently are not particularly choosy. This is also confirmed by examination of the stomach contents of certain species. The available supply of coral polyps and other micro-organisms is so huge that the individual species do not necessarily have to enter into competition. I did note, however, that the fishes showed markedly different degrees of specialization in their choice of dwelling-place. In total I made 221 dives down to a maximum depth of 90 metres (300 feet), to study the natural distribution of these fishes. Most of them are found only in shallow water down to a depth of some 20 metres (65 feet). Some, like the butterflyfish *Heniochus acuminatus,* I observed down to as far as 60 metres (almost 200 feet). The crew of a Cousteau submarine observed *Chaetodon falcifer* at a depth of 150 metres (almost 500 feet). I also detected specializations in particular habitats. For instance, the butterfly-fish *Megaprotodon trifascialis* was always found in the immediate vicinity of Acropora coral, whereas the particularly aggressive, dirty-yellow *Chaetodon kleinii* predominantly occurs in rocky areas, frequently even such areas quite free from coral.

These preferences for certain ecological niches enable the butterflyfishes to achieve maximum utilization of the facilities offered by living conditions, as well as peaceful coexistence of the species, since such preferences prevent competition. In order to discover how many species of butterflyfish could occur in a reef region, I totted up the species observed as I swam over a straight stretch of 1 to 1½ kilometres (about ¾ mile). I also estimated the size of fish and wrote down on my underwater note block whether they were living alone, in pairs or in groups. I remained in the same area of reef, since the mix of species can vary from one zone to another. At Eilat on the Red Sea I recorded eight species, and on several reefs in the Indian Ocean I even counted as many as fourteen different species. These counts confirm at the same time that the majority of butterflyfishes pair strictly for life. The very close proximity of so many closely related fish could lead to their cross-mating. Cross-breeds are generally infertile or defective in some other way. As a rule they are also less well adapted and are therefore selected out under the pressure of competition with the 'pure' species. Crosses between different species are also ineffective from two points of view: they involve a loss of time and of the biologically valuable germ cells, while in addition the specialization already achieved is being put at risk. How then are these interspecific crosses prevented?

Brilliant colours and pairing for life

Nature would appear to have solved this problem by means of brilliant, poster-like colours. The use of such colours in those families of fish which comprise numerous species serves to distinguish between the specialists with their varied adaptation. They can, however, also at the same time act as distinctive markings peculiar to the species, so eliminating cross-mating. The greater the difference between the coloration and form of two different species, the easier it is for them to maintain the proper distinction. This makes it easy to understand why nature has assigned such a heavy degree of selection pressure to visual distinctions. A further safeguard against cross-breeding is the principle of mating for life.

Wickler has demonstrated that monogamy and reproduction, pair formation and parental care are not necessarily linked with one another. The phenomenon of pairing for life was not necessarily 'discovered' by nature for the purpose of producing and raising young. Our observations to date show that the butterflyfishes which pair for life do not raise their young.

If the choice of partner is undertaken with great care and the pair formed then swim around in company for the rest of their lives, the likelihood of faulty crossing is appreciably reduced.

Fish which are ready for mating will take a partner most closely resembling their own species of they find no fellow-member of the species to pair off with. I have actually observed on a few occasions pairs made up of two different species. On one reef off Madagascar I found a butterflyfish pair consisting of one *Chaetodon auriga* and one *Chaetodon falcula*. The partners were of equal size and were similar in pattern and colouring. In the Gulf of Aqaba, however, I encountered a most ill-matched pair: a rare fish for that region, a lemonfish *(Chaetodon semilarvatus)* swimming round with a butterflyfish *(Heniochus)*, with which all it had in common was its size and its flattened body shape. The patterning and coloration of the two species are quite different. As a means of guarding against faulty crossing, butterflyfish patrol the boundaries of their distribution areas more frequently as a pair than in densely-populated regions, where it is perfectly possible to see solitary individuals. The fact of mating for life is thus particularly a means of guarding against interspecific pairing in thinly populated areas. Brilliant colours as species recognition signals, and the habit of mating for life, help to maintain the special characteristics of the species and preserve the specialization which has been achieved. That different types of adaptation also favour differing coloration is clearly shown by the young of many species of reef fish.

For the young of all the brilliantly coloured fishes there is the difficulty of holding their own against the adult members of their own species. How would they be able to reach maturity if they were always being attacked by their elders?

Here too, nature has found a number of ways out of the difficulty. Some of the juveniles live in zones where adults of their species do not occur. For instance, they settle in tidal pools. But if they do live in the areas which are also inhabited by adults, they seek out the places where they will be secure from their elders' attacks. Many take up station between the spines of the large diadem sea-urchin or between the small branchlets of the Acropora coral. Not only do the young fish live in another region, but they also probably feed on other food sources. In addition to this they are differently coloured. Many juvenile fishes differ so much in patterning and coloration from the adults, that even fish experts can make mistakes and describe some of the juvenile forms as adults. Only as their development progresses do they undergo a colour change and slowly assume the full colour pattern of the adult fish (Plates 50–53).

This adoption of false colours makes it possible to achieve peaceful coexistence of old and young. Even in the aquarium, the differently-coloured age groups rarely attack each other. The young fish frequently bear a striking round component in the body design – an eye-spot which is usually lacking in the adult fish. Presumably this is intended to reinforce the effect of masking the species-specific characteristics; eye-spots are found particularly on such young fish as have a body pattern which is already fairly similar to that of the older fish.

The language of signals

As we swim across a reef landscape, we soon observe that the fish are by no means dumb creatures. All around we can hear churring, clicking or booming noises, and we cannot conceal our surprise as we look around for the mysterious creatures emitting these sounds. To whom are these acoustic signals directed? In addition, behaviour patterns are often demonstrated so strikingly that we find ourselves very anxious to learn the significance of the movements. In such cases I have frequently searched the surrounding area for a possible receiver, to whom the behavioural signal might be addressed, since it might very well betray by its reaction whether it had understood the signal.

Communication in the undersea world is just as multifarious as animal communication on land. It is, of course, true that investigating it is an activity which encounters considerable difficulties. In the well-lit coral reefs, especially in shallow water, the language of colours and patterns plays an important part, as we have already seen in discussing the brilliant poster colours of many fishes.

Signals between animals can comprise coloration and patterning, sounds, smells, movements and tactile stimuli. Intraspecific or interspecific communication always requires one signal emitter, which sends out a message, and one receiver, which understands the signal. But how do the creatures know the meaning of the individual signals? Do they possess an innate programme for the purpose or do they have to learn how to decipher the code? The meaning of signals is frequently only learned by the receiver in the light of experience, but it is generally inborn mechanisms which cause an animal to have any reaction to a signal at all. However, there are certain cases where we know the meaning to be assigned to signals used in communication between fishes, and the methods of information transmission used by the various species.

Mating signals

For some years now I have been observing the domino damsel *(Dascyllus trimaculatus)*. This restricted-range fish, some 10–13 centimetres (4–5 inches) in length, lives in groups which can comprise up to eighty individuals. At certain intervals, the entire colony spawns. Shortly beforehand the males prepare an area and set up a nest site in it. Suddenly they begin to leap about in a peculiar manner; they shoot steeply upwards, tip over after having covered about three feet and return to the nest site. This process is repeated many times. Throughout the colony we can observe males suddenly springing in this way and signalling to their partners by means of this striking behaviour. The message is: this is my district, which I shall defend against all invaders! Presumably the signal also serves at the same time to stimulate the females.

On the day of spawning the males begin intensive springing behaviour shortly before sunrise. It is now possible to hear quite clearly a powerful churring noise – *brrr* – which is emitted each time the fish reaches the top point of its leap. The females, ready to spawn, have already taken up position in the vicinity of the nest site. Once a male has finally succeeded in attracting a female by these signal leaps, they swim together to the site. On the way there the male darts alternately to right and left, and only shortly before they reach the nest do the fish swim together as a pair. As soon as she arrives at the nest site, the female immediately begins spawning. The male swims to and fro several times over the fresh-laid spawn, expelling its sperm, and watching its partner's spawning activity in the interval. During this process the forepart of its body trembles to and fro in an obvious manner and, once more, the fish emits a clearly audible *brrr* sound.

Frequently the female is still busy spawning while the male has already attracted other females by means of his signal leaps. Once more the male accompanies a female who is willing to spawn to the spawning place, the two fishes swimming as a pair. However, if it catches sight of the earlier partner still occupied with spawning, the male drives away the 'new love' and returns to the first female. During these observations of the domino damselfish it became clear that the males and females of the colony who did not know each other previously have to adjust to each other, that is to attain the same degree of physiological preparedness, in order to achieve synchronous emission of sperm and spawn. By means of the visually striking signal jumps and the clearly audible churring noises, the males induce the appropriate state of readiness in the females of the colony, so initiating egg ripening and subsequent laying of the eggs. Only recently was it shown by research workers at the University of Tel Aviv, in the aquarium, that the female damselfish can be induced to spawn without the presence of the male, simply by hearing the male calls transmitted through the water. This is a clear indication of how important sound communication can be even under water.

Fighting signals

Anne Rasa of our Institute observed with another damselfish *(Pomacentrus jenkinsi)* that when conflict arises it 'speaks' to its opponent by the colour and patterning of the eye. If a transverse bar appears in the eye, the fish is in a particularly aggressive state. At the same time, there is a relationship between the eye colour and pattern and the position of the dorsal fins. For example, preparedness for flight is indicated by raising the fins. Even the human observer can predict in the light of this how a fight will finish, if he can interpret all these characteristic signs. It is true that in order to do so he must look precisely into the eye of the fish, something which is not always easy to achieve, because the colour and patterning of the eye are not so evident as are other colour features on the body.

Eye-spots

Some fish have similar methods of indicating their attitude in the event of a fight. Particularly with butterflyfish we observe large, generally black, signal spots at the rear of the body or on the fins, which are known as eye-spots. The true eyes of these fish are generally camouflaged by a broad black stripe. It used to be thought that the eye-spot was intended to produce the illusion of a false head. Many predators do in fact attack the head of their prey, 'homing in' on the eye. If the intended victim were now to swim off suddenly in the opposite direction, this would confuse the predator.

Wickler established that the attacks of the small predator blenny *Runula,* which bites out pieces of skin from other fish, are directed towards the eye-spots. The camouflaged eyes themselves are generally left untouched. These predators, however, are not so common that their possible victims would need to assume camouflage stripes across the eye and false eyes on some other part of the body, for fear of being attacked. I have studied in the aquarium the way in which eye-spots also serve as intraspecific signals. They serve an attacking rival as a target. The perceptive system is strongly attracted by a striking, round component of a pattern, such as the eye-spot. A rival would in the heat of its aggression bite at those points of the opponent's body which make a particular visual impact on it. Certain butterflyfishes have additionally developed the capacity to change the colour of their eye-spot; the long-nosed butterflyfish *(Chelmon rostratus)* and the four-eyed butterflyfish *(Chaetodon capistratus)* are instances.

Quite by chance I was able to observe the strongly ritualized threat displays of these species, which proceeded like jousts governed by strict rules. Opponents of equivalent strength threatened each other broadside on, trying meanwhile to aim at the eye-simulating spot. The fish arrange themselves head to tail, so that the eye-spot of the one is fully in the other's vision. As soon as one of the two feels itself dominated, it suddenly changes the colour of the spot and folds down its fins. This is the sign that it abandons the fight. A life-and-death struggle is therefore avoided by the use of such messages. The fact that threat displays of this kind in the aquarium frequently progress to fights with fatal consequences is attributable to the confined space in which the fish are imprisoned. In these cramped conditions it is in fact impossible for the loser to escape, so that he is hounded to death by the victor. In the free spaces of the coral reef such situations cannot ever arise, because ample facilities for flight or for hiding are available. Acting as 'mood indicators', the eye-spots signal the opponent's readiness or otherwise for battle. In this way nature prevents senseless shedding of blood – in a manner from which *Homo sapiens* might well learn.

Threat signals

Signals can also frighten off rivals or predatory enemies. Creatures which employ signals specially directed towards those who might wish to eat them are generally poisonous as well. In a similar way, our showily-coloured wasps leave behind them with their stings such a durable impression on those who threaten them that they are very likely to leave other wasps at peace thereafter! The firefish or long-spined turkey-fish, when threatened with danger, spreads out and stiffens its fins to transform itself into a most impressive form (Plates 109, 110). The poisonous spines on the dorsal fins sway slowly to and fro, and anyone who has once come into contact with this fish will avoid any such contact in future.

The coloration of the anemonefish has also been explained as being a warning signal (Plate 64). Should a predatory fish once touch the tentacles of the anemone, it will be badly stung, and will avoid both the sea anemone and its inhabitants, the anemone-fish, *Amphiprion*, which it will recognize subsequently by their showy colours. Some of the species of *Haemulon*, the grunts, have a particularly magnificently coloured interior to their mouths, and in order to frighten off attackers they open their mouths wide. I have observed similar behaviour in a moray eel *(Lycodontis)* in the Red Sea. Every time I have swum into the territory of this moray eel, it opened its mouth wide, displaying the showy bright orange-red interior (Plate 65). When the eel was attacked on one occasion by a school of damselfish, it tried to frighten off its assailants in the same way.

Social signals

Another type of signal functions like a club badge, to indicate membership of a clearly defined group. Many freshwater fish assume a particular shoal costume, which they show only when swimming in groups. *Pristella,* for instance, bears a conspicuous marking on its dorsal fins. Experiments have shown that fish maintained in solitary confinement preferentially swim after members of the group which bear this marking, rather than after those from which the marking has been surgically removed; thus the pattern in question must serve as a 'follow-me' sign.

The domino damsel *(Dascyllus trimaculatus),* already referred to, carries in the juvenile stage large white patches on forehead and flanks, which frequently become smaller on adult fishes or disappear altogether. The young fish, on which the patches are particularly large and conspicuous, swim together in dense schools. I therefore assume that these markings serve as a social symbol, and a recognition signal among the younger generation (Plate 64). Eibl-Eibesfeldt writes that many of the sociable fish which swim around in the open sea have dark patches on the tips of the tail, or other simple markings which stimulate the crowd to follow them (Plate 66). The easiest instances in which a social function of patterns in fish can be proved are those where there is a faculty for colour change and the signal patches are exhibited only in certain situations. Thus the large batfish *(Platax),* which can be found either as solitary individuals or in groups, displays a large black patch on the ventral surface, which is otherwise brightly shining silver, as soon as it swims in a group.

Nocturnal flashing signals

One moonless September night I saw in the coral reef near Ras Burga on the coast of the Sinai Peninsula a bright flash. A fiery ball appeared to glide slowly along the edge of the reef. We immediately dived into the water and discovered a school of fish holding a nocturnal firework display of flashes. These fishes were members of the genus *Photoblepharon* (Plate 62). Underneath the eye they have a large white photophore or luminous organ, which they can brighten or dim like a searchlight. The bright patch of light is extinguished by drawing a pigmented membrane across the organ. These fish emit light signals, but to whom? In one of the females which we captured we discovered a large number of eggs, so it was clear that the spawning time was not far off. And during the following nights the fish were observed swimming together in pairs and 'flashing' vigorously. When we started following the pairs, they all returned to their groups. Moreover, the members of the group were 'flashing' to each other. In consequence we concluded that these light signals serve to keep the school of fish together or to enable one individual to locate its mate.

Communication between cleaner and customer

I have already described the specialization of the cleaner fishes which collect the parasites from the bodies of other fish. This cleaning industry is very widespread along the reef. Young fish in particular seize the opportunity to look for food on the body surfaces of other fish. Some species have become 'professional' cleaners: in the Indo-Pacific region, the Pacific wrasse or cleaner wrasse *(Labroides dimidiatus)*, in certain areas of the Pacific, *Labroides phthirophagus* and in the Caribbean the tiny cleaner goby *(Elacatinus oceanops)*. Shrimps, too, crawl around on the bodies of fish and busily collect parasites with their tiny claws (Plate 69). I shall examine in more detail the significance of these cleaners for the fish population of the reef.

The cleaner fishes' clients take up a very clear invitation position, to indicate that they wish to be cleaned (Plates 100, 103). These behavioural signals are discerned and understood by the cleaners even at considerable distances and they swim up to their customers. But how does the client recognize the cleaner? Most of the 'professional cleaners' wear a uniform of gleaming blue longitudinal stripes, which are easily seen and can be taken as a signal to the would-be customer. In addition the 'professional cleaners' first of all perform a conspicuous to-and-fro dancing movement, especially before unknown clients. Cleaner shrimps draw attention to themselves by agitating their long white antennae and by making short forward-and-backward movements with their bodies. These signals show the client that he is welcome (Plate 68).

Involuntarily the observer fears for the life of the cleaner, as it swims unperturbed into the gaping jaws of the customers – even predatory fish such as groupers or barracudas – as it were directly into the lion's mouth (Plate 67). When they clean the gills, the cleaners disappear entirely behind the large gill covers. Once the client feels that he has been thoroughly treated, he invites the cleaner to leave his gills by closing the gill cover once or twice. On receiving this signal the cleaner stops work; communication has taken place between the cleaner and the customer. Thus behavioural patterns in both cleaners and customers have come to form part of interspecific communication. In addition they recognize each other by visual signals which they immediately classify as 'friendly' on the basis of previous experience.

But how did this communication between the two partners develop? We must assume that much of the activity in the cleaning industry is acquired, that is obtained by experience in the course of the development of each individual. It is virtually inconceivable that a grouper could be born with a system of recognition patterns for all the cleaners occurring in the reef; it would be equally impossible for the cleaners to recognize all their clients 'instinctively'. On the other hand, the professional cleaners presumably do possess an innate pattern for recognizing the cleaning invitation position of their clients – a position which has few distinctive features – and this recognition pattern stimulates them to swim towards the potential client.

Warning – cleaner mimics!

Some fish exploit the friendly signals employed by the cleaners. The sawtooth blenny *(Aspidontus taeniatus)* imitates the cleaner in all external features and in behaviour so skilfully that even the trained eye of the marine biologist is deceived. The cleaner mimic has a powerful pair of jaws and approaches the client 'disguised' as a cleaner, so that the customer fish promptly assumes the cleaning invitation position. But this wolf in sheep's clothing is not a hard-working collector of parasites, but a predator. Using its razor-sharp front teeth with lightning rapidity it rips out small pieces of the client's skin. Only then does the client see through the deception and drive the presumed cleaner away.

In the Red Sea I was able to observe a different kind of cleaner mimic. Young blennies of a species of *Runula* imitate young *Labroides dimidiatus,* which wear a colour pattern different from that of the adult fish (Plate 70). The young *Runula,* finding it easy to approach the clients because of this deceptive imitation of the colour signals, starts biting the unsuspecting fish, preferentially attacking their eyes. When on one occasion I observed the attacks of the cleaner mimic in two different, thickly populated regions of the reef and counted the total number of incidents over a given period of time, I discovered an interesting piece of behaviour. In the sparsely populated region the cleaner mimic bit only at passing fishes and avoided all the home-range fishes, which in fact on a strikingly large number of occasions chased it away. Clearly these local residents had in the meantime recognized the cleaner mimic by certain distinctive characteristics. On the other hand, in the densely populated region the clients are unable to learn to recognize the skilful impostor very quickly, because they encounter it less frequently. Since a large number of client fish come to each cleaner, there is no great likelihood that the cleaner mimic and the client will meet each other very often.

In this way communication by means of signals is exploited 'parasitically' by other fish. The more refined the imitation of the signals, the more easily deceived are those for whom the genuine signals are intended.

71 *(Opposite)* Methods of camouflage. The jewfish *(Epinephelus itajara)* breaks up the outline of its body by a pattern of dots and stripes. This camouflage pattern is particularly effective at long distances.

73

74

◁72 The Wobbegong shark *(Orectolobus ogilbyi)* uses a camouflage pattern which disrupts its entire body shape, so that the body of the shark melts into the background.

73 The scorpionfish *(Scorpaenopsis gibbosa)* is exceptionally well camouflaged and lies motionless in wait for its prey, which it captures with combined suck-and-snap movements. The sting from its poisonous spines, carried on the dorsal fin, can be fatal even to human beings.

74 The rabbitfish *(Siganus)* lie motionless on the sea-bottom at night and have a diffused patterning on their bodies to camouflage them.

◁75 Many sea-dwellers absorb colouring materials in the course of eating. A small snail *(Neosimnia)*, lives on the magnificently coloured gorgonians and arrays itself in colours similar to those of its host. It even imitates the colour tones of the coral polyps, so as to correspond better to the background against which it occurs. This camouflage technique deceives the creatures which might want to eat it, since they are not interested in coral.

76, 77 It is only a few years ago that the association between the whip corals *(Cirrhipathes)* and small gobies *(Cottogobius)* was discovered. Whatever the colour of the coral to which it clings, the fish adopts the colour of its host. It eats pieces from the coral surface and also picks off a rich yield of plankton, which the coral polyps themselves have trapped. As in Plate 75 with the snail, so in this instance also, the fish imitates both the colour and the contours of the coral polyps.

78 The scribbled filefish *(Osbeckia scripta)* from the Red Sea camouflages itself by means of colour changes. Pigment cells in its skin enable it to imitate the colour of the background.

79 The Caribbean trumpetfish *(Aulostomus maculatus)* hides between the branches of horn coral. It modifies its colour to suit its background and orientates its body in the same direction as the coral branches. This camouflage deceives not only the predators, but also the fish on which the trumpetfish preys, which swim unsuspectingly too near.

80 Garfish and halfbeaks *(Strongylura* and *Hemiramphus)* hunt close below the surface of the sea. The counter-shading of their bodies makes their coloration blend into the background of water.

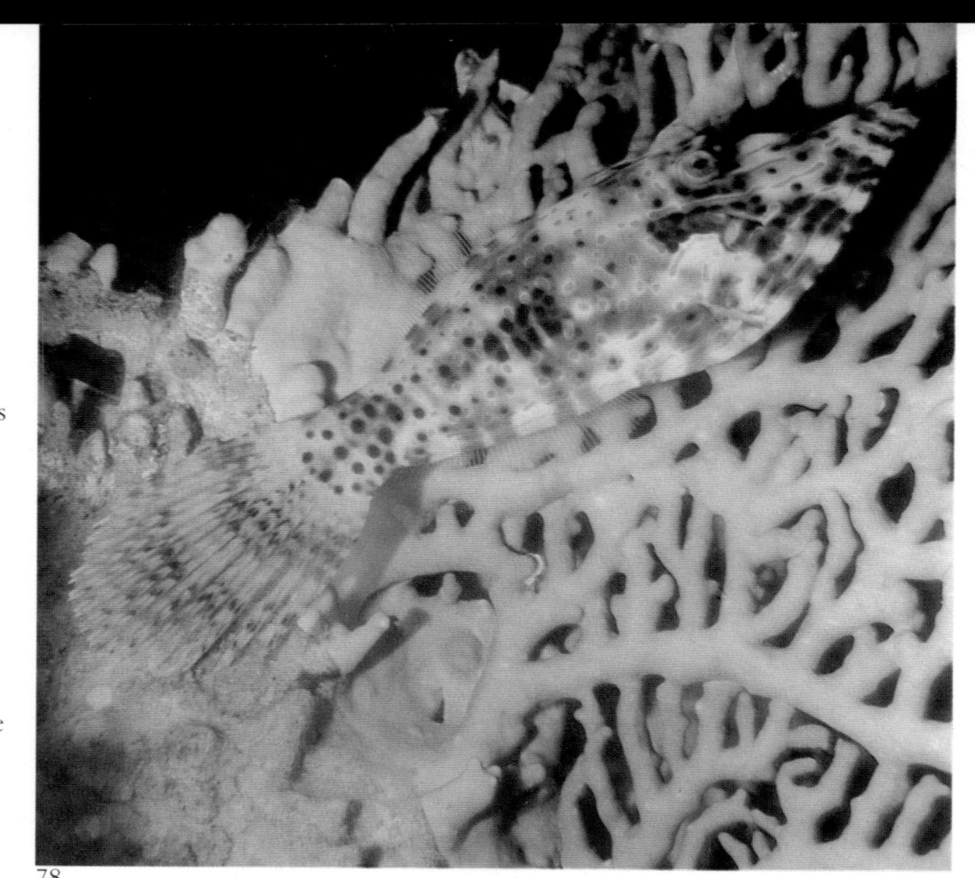

78

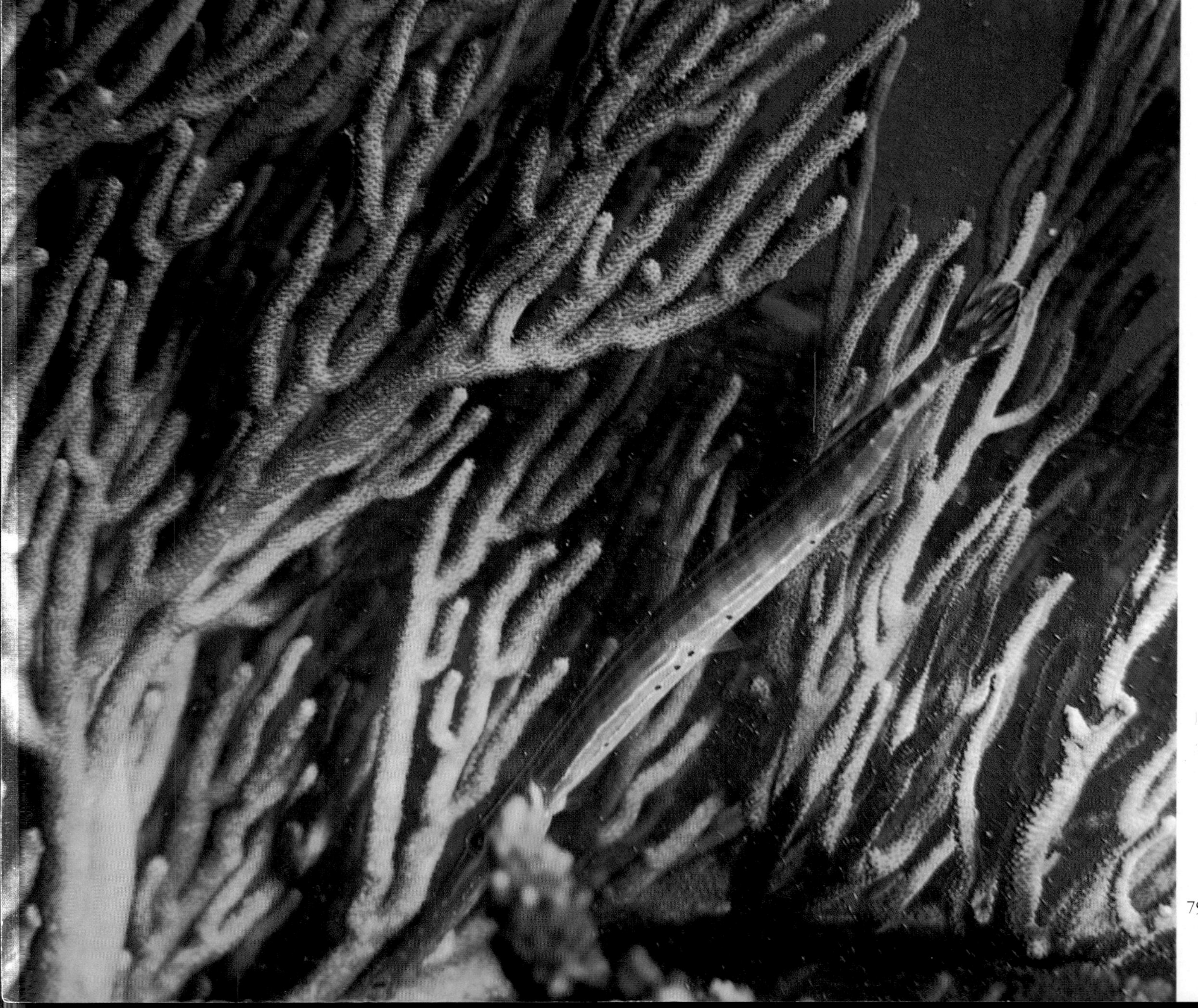

79

80

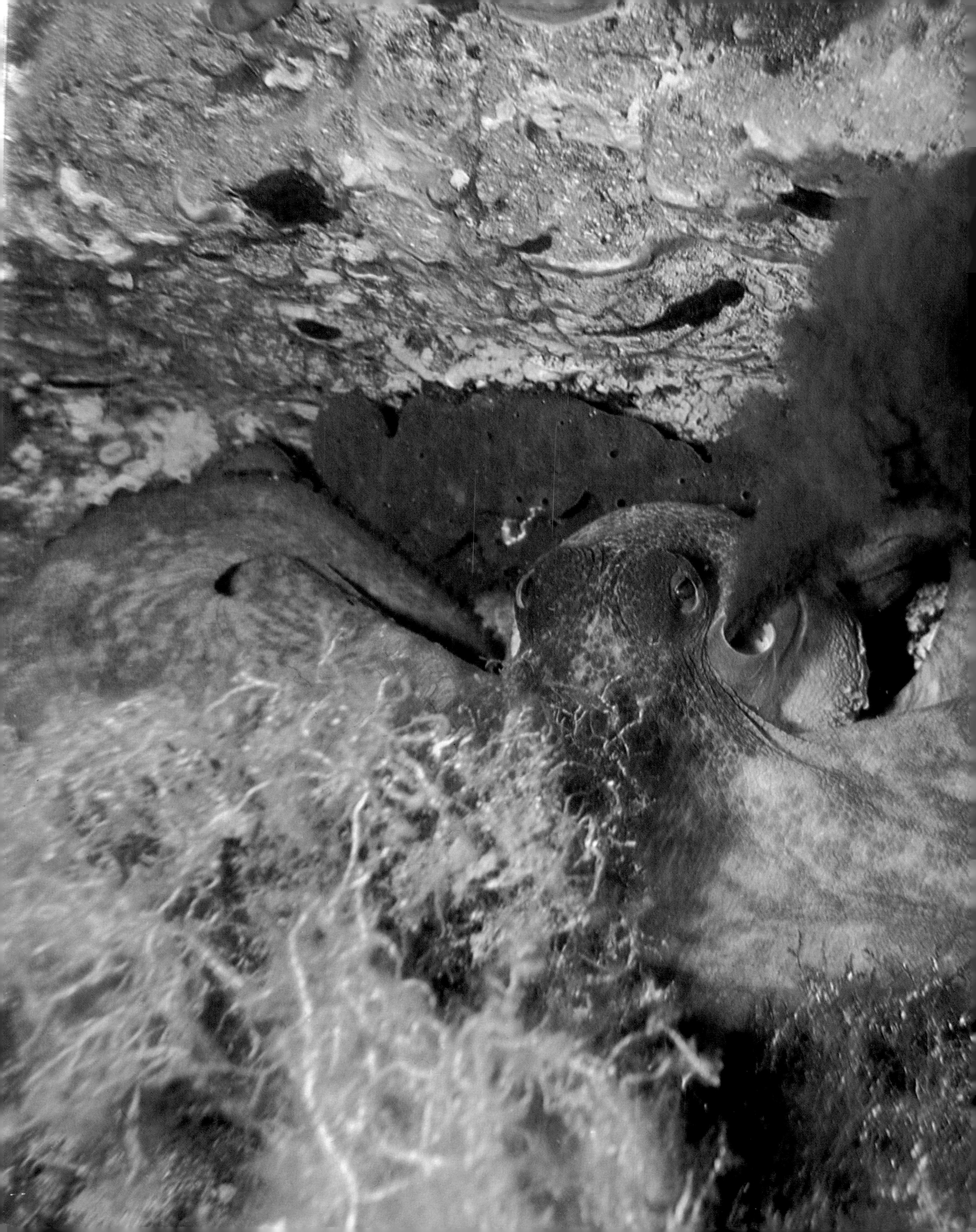

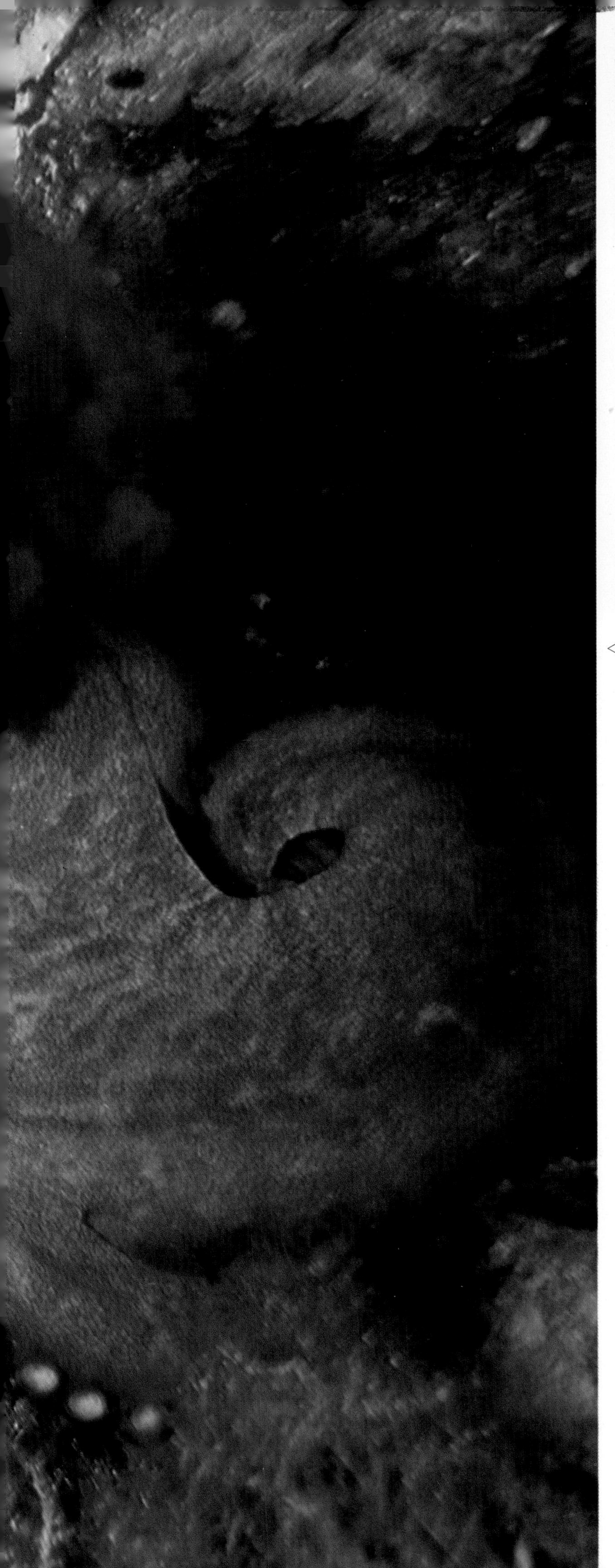

82

◁81 The octopus lives in fissured reefs and by day occupies a cave, which offers it protection and which it defends vigorously against members of its own species. It heaps up stones, snails and bivalves at the entry to its cave. In the event of danger the octopus can expel water from a funnel-shaped jet in its mantle and thus uses a reaction thrust to flee, using the same principle as a rocket. The octopus is a master of camouflage. Waves of colour move over its body as it steals across the bottom on its tentacles. It can assume a colour which matches that of its background with extreme rapidity. In addition, fringelike processes can be extended above its body surface, resembling in appearance the structure of the ground around it. If it is badly disturbed, the octopus can emit a cloud of ink, as shown in the photograph. This cloud is thought to interfere with the olfactory powers of the predators and to hide the octopus.

82 Although the octopus and the cuttlefish *(Sepia)* are related to the snails and bivalves, they have highly developed eyes which are functionally similar to those of the vertebrates. Frequently it is only the huge eye of the camouflaged *Sepia* which remains visible.

83 Brood care. The snail *Crimora popillate* packs its eggs into a gelatinous protective egg-capsule.

84 *Amphiprion bicinctus,* the anemone-fish, fans and guards its eggs for up to ten days, where they have been laid in the shelter of the anemone. Often the spawn mass is eaten by small crabs.

85 The triggerfish *Balistes fuscus* lays its glassy, transparent eggs in a pit in the sand and defends them vigorously. In these circumstances even divers can be attacked and bitten. But this brood care only lasts a single day.

86 *(Overleaf)* A small octopus *(Octopus aegina)* lays its eggs in a heart cockle, where it guards them from enemies.

83

84

3 Camouflage, mating and brood care

I am slowly swimming down the sandy slope of the southern end of the small island of Nosy Tanikely, off Madagascar. The undulating lattice of sunlight dances jerkily across the sea-bed like patterns in a kaleidoscope and, when the light falls at a certain angle, creates rainbows on the sand. A few sand eels dive head-first away from me into the sandy bottom. The invisible pull of the undersea current transports me across a forest of acacia-like algae, which grow in their thousands on the sandy slope below me. From my 'bird's-eye view' the bottom of the sea appears like an African savannah.

Suddenly this somewhat monotonous scene springs to life. A colony of *Astropyga* sea-urchins is moving across the sea-bed. Their pretty dark-violet bodies stand out boldly against the bright background. It is advisable to take care, since I have often been stung before by the pointed, venomous spines of this urchin. With one powerful stroke I swim on. And then I observe something quite astonishing. Obviously frightened by my abrupt movement, one of the urchins takes to flight and breaks up into hundreds of small particles. Fascinated by this sight, I watch it closely and see how these little bodies whirl freely around in the water and then swim in a group towards a sea-urchin. In a flash, they have surrounded it in a dense swarm and it is almost impossible to see them. In shape and colour they have once again assumed the form of the sea-urchin.

Even the trained eye of the diver finds it difficult to discern that what he has before him is a swarm of fish trying to camouflage itself against the spiny background of an urchin – a piece of collective camouflage. The tiny cardinalfish *(Siphamia argentea)* has hit upon an extraordinary way of hiding from its enemies. The little fish crowd closely together and increase the apparent size of the sea-urchin. If I now look very closely, I can distinguish the shape of each tiny fish. But their imitation of the urchin is so successful that many predatory enemies of the *Siphamia* will be deceived.

The hunters and the hunted live in the reef as close neighbours, in a manner not found in any other association of living creatures. Evolution has taught the hunted to avoid contact with the enemy. One way in which they can do this is by camouflage, making themselves as invisible as possible to the predator. Conversely, the predators too have developed the same device to help them to approach their prey more rapidly and inconspicuously.

Many of the camouflage methods used in the coral reef are similar to those which have been proved so successful on land. The standard camouflage patterns can be usefully applied anywhere, quite independently of the evolutionary development of the creatures themselves – even in modern warfare between human beings.

Body colour as a means of camouflage

A fish is particularly easy to see if it throws a clearly visible shadow and if, in addition, it differs considerably in body coloration from its surroundings. In this case the fish can best camouflage itself by producing the negative of these characteristics, in other words, by avoiding an obvious shadow and by adopting colouring which is similar to that of the surroundings. Thus it is that insects on trees or grassland are often green in colour, whereas animals inhabiting the snow-covered landscapes of the far north are frequently white.

In the coral reef, similarity of colour between the creatures and their surroundings frequently occurs with those which occupy extreme ecological niches. Years ago I caught the bright-red sea-slug *Hexabranchus* in the surf of the Red Sea. In the aquarium this creature of rare beauty swam through the water with peculiar jerky movements of its body axis. And then I noticed an equally red shrimp *(Periclimenes imperator),* which had fallen from the back of its host as a result of the latter's swimming movements and by chance found itself alone in the water. This shrimp can also live in the mass of tentacles of a sea-cucumber, where it steals the food from its host. In this instance too the shrimp is difficult to detect, because it is able to adopt a coloration similar to that of the brownish tentacles of the sea-cucumber.

Another example of colour similarity comes from the Indian Ocean. The small goby (*Cottogobius,* Plates 76, 77) adopts the colouring, tint for tint, of a yellowish whip coral *(Cirrhipathes).* The fish is specialized to collect the particles of food which the coral polyps laboriously trap for themselves. This camouflage behaviour is of great importance for the goby, because the thin branchlets of the coral stretch far out into the water and a fish of a different colour would be readily visible to its enemies. In the Red Sea the same goby lives on a greenish whip coral and has here 'dressed' itself to suit.

Many snails live on a very wide range of host creatures and absorb with their food the pigments of the host. Plate 75 shows the snail *Neosimnia* on a reddish gorgonian. The colours of the snail and its host match exceedingly well.

Even the absence of colour can be an effective protection against predators. The plankton organisms are generally translucent and are therefore not easily visible in open water. Shrimps are often so transparent that it is difficult to discern their bodies. The shrimp *Periclimenes affinis* lives on the sea anemone *Discosoma.* The shrimp is transparent, but remarkably enough possesses a white eye stripe and a striking patch at the end of its tail. Its transparency enables it to become almost invisible on the tentacles of the anemone, and even the bright spots of colour blend into the coloured background. Only when the shrimp swims off into free water is it possible to see the blotches of colour which stand out on its body. Presumably these blotches serve as cleaner signals, since many species of *Periclimenes* perform this task.

Colour changes

One of the most important ways of assimilating body colour to the background is to make use of reflex colour change. Most creatures capable of undergoing colour changes are provided with complicated cell structures in the skin. These colour cells consist of a cellular body which extends into numerous processes. The pigments are contained in the cell plasma. There are special colour cells for each colour, the cells being superimposed in layers. When the pigment in a cell spreads out into the finest extensions of the cell, the colour becomes clearly visible. The cell processes serve to enlarge the surface, so that the pigments are more readily visible. Nerves or hormones trigger off the colour changes in the pigment cells.

Cuttlefish are real masters of transformation. I have frequently watched the small ten-armed *Sepia.* On the meadows of eel-grass *(Zostera)* they are green, but on touching down on the sand they immediately adopt its colour. Before they land, waves of colour pulses sweep across their body surface until the cuttlefish settles on the sea-bed like a helicopter. Suddenly they agitate their marginal fins and throw sand over their bodies, so perfecting their camouflage. Only the huge eyes now peep out of the sand. Even the eight-armed octopus can camouflage itself in a second by colour change (Plates 81, 82). In addition, the octopus can erect fringe-like processes on its body surface, thus enabling the creature to adapt the outline of its body very closely to its background.

Those creatures which inhabit areas where there is no cover can change their colour with astounding rapidity. Flounders, soles and plaice become completely invisible. However, camouflage does have one disadvantage within a species, namely that the members of the same species are also deceived and communication with them is thereby made more difficult. For this reason certain fish have developed eye-catching colour signals which they display on their pectoral fins, which can be rapidly erected, so that they indicate their presence to other members of their species.

On the sandy bottom of the Indo-Pacific region lives the dangerous lizardfish *Synodus,* which eats a great many small fish. It has developed a refined system of tactics to approach its prey. It creeps towards its victim, secure in its sand-colour camouflage, and then leaps on it with extreme rapidity. I once set up a glass jar containing young garden eels on the sand and watched the movements of a lizardfish carefully stalking its prey. As it made its habitual pounce, it struck the outside of the glass jar with all its force; both the predator and its intended victim were equally puzzled by this, with the result that they panicked. Here was a clear proof of the advantage which can be drawn by a predator from the use of camouflage.

Considerable skill in changing colour is also exhibited by defenceless creatures which inhabit a coloured or highly structured background. The trumpetfish can change its colour within a few seconds and, in addition, camouflage its body shape, by imitating the colour of certain sections of the background so skilfully that it can hardly be discerned (Plate 79).

Disruptive patterns

A uniformly coloured body occupies a given area in our field of view, standing out from the surroundings with a certain degree of contrast. This is why we are able to recognize many animals by their shape at considerable distances. The perceptory system 'prefers' the symmetry of figures and is readily led to follow their curves. This specific feature of perception is utilized in one form of camouflage. The creatures' bodies are marked with irregular stripes, patterns or blotches, which disguise the shape – they confuse the observer's eye so that he cannot clearly see the contours.

The grouper (*Epinephelus itajara,* Plate 71) is stippled all over with small black spots and in addition has irregular white stripes on its flanks. These patterns disrupt the outline of the predator. The Wobbegong shark *(Orectolobus ogilbyi),* photographed in Australia, is exceedingly well camouflaged by a disruptive pattern (Plate 72). The three rabbitfishes (*Siganus,* Plate 74) were photographed at night. Their diffusely striped 'nightshirts' hide them from the eyes of predators, since these hunt for preference at twilight. Night colorations adopted by fish are therefore frequently intensified in many instances during twilight, because the danger from their enemies is particularly great at this time.

The Cambridge University zoologist Hugh Bamford Cott has shown that when a zebra-striped black-and-white damselfish *(Dascyllus aruanus)* is in front of a dark background, only its lighter parts are visible, and the form of the fish is disrupted. In the vicinity of light-coloured corals, the white body stripes prevent clear recognition of the body contours. The same is true of the butterflyfish *Heniochus,* whose contrasting black-and-white striping is very clearly visible at close quarters, but is hardly detectable against a background which is alternately light and dark.

Probably all these patterns take effect only at fairly considerable distances, the form remaining clearly recognizable from close at hand. The striping of the black-and-white damsel then even becomes an attractive signal for others of the same species.

With many butterflyfish the eye is marked by a conspicuous black stripe. Is this intended to frighten the predator, if it does not see the fish's eye and if its shape appears unfamiliar, or does the eye stripe act as a camouflage to mask the forepart of the fish's body? This is a question we cannot answer without experiment. All that we do know is that the eye of a predatory fish obeys the same laws of vision as the human eye; nonetheless, we must be prudent in seeking to explain this kind of camouflage device, because we do not know which features are important or unimportant to the predator.

The interplay of shading and countershading

However well adapted in colour a creature is to its background or however efficiently camouflaged by other means, its shadow betrays its presence and attracts the attention of its enemies. Many creatures have consequently developed methods of avoiding casting shadows of their bodies. When at rest, butterflies turn their folded wings edgewise towards the sun. In this way the shadow falling on the ground is only a narrow band, and the butterfly is thereby rendered less conspicuous. On the sea-bottom too it is easy to see how the creatures avoid creating a shadow which would be clearly visible. I have often seen crabs running first right and then left in their flight; suddenly they will press themselves flat against the bottom, so that their shadow disappears completely. Generally, in addition to this, the crab's back is camouflage-coloured, so that the efficacy of this escape manoeuvre is increased. I have observed this behaviour particularly well in the case of the ghost crabs *(Ocypode saratan)*. The outer covering is sand-coloured, so that the crab can hardly be distinguished from the sea-bottom.

Only in shallow water will a predator's attention be caught by the shadow of a possible victim. I once witnessed a dramatic hunt between a razor lipfish and a small flounder. The hunt began when the predator saw the flounder's shadow; from above the flatfish had been very effectively camouflaged by its coloration and yet the lipfish 'knew' where to poke into the sand to chase the flounder out. Each time the flounder resorted to flight, it was betrayed by its shadow. And nevertheless it managed to get away. Cuttlefish *(Sepia)*, too, duck down to the sea-bed and so avoid casting shadows. This avoidance of a conspicuous shadow is not the only element in the camouflage of most creatures. Frequently the body is also adapted to its surroundings by colour change or by some other optical camouflage device.

In many habitats, the prevailing conditions of lighting do not produce heavy cast shadows – for instance, in a dense forest of algae or in the wide spaces of the open ocean – but even so the light can betray a creature's position. Since the illumination comes from above, a fish's back is more strongly lit than its belly, so that the body stands out against the background. To avoid being shown up in this way, many of the creatures adopt their own shadow camouflage, applying dark patterns or pigments where the effect of the light is most marked, i. e. on their backs which are turned to the direction from which the light comes. The back is darker-coloured, and the belly white, producing as it were a negative of the natural body shadow. The final result is that the body appears to be flat and its three-dimensional effect is reduced. The garfish which hunt close beneath the water surface are so well camouflaged by countershading that they completely melt into their background (Plate 80). *Caesio* fish live in great shoals in the open water above the reef. They are steel-blue on the back and whitish on the belly. At considerable distances, when viewed from the side, they merge with the blue background of the water. Even from above they are difficult to detect. However, if we observe them from below, their bodies show up as a black line against the bright mirror of the water surface. This observation has brought a number of research workers to doubt the effectiveness of countershading. It cannot be denied that countershading exerts its camouflage effect primarily from the side and from above, but this alone is an important factor for the individual and contributes to the survival of the species.

The American specialist on coral reefs, William H. Longley, has written that countershading is almost universal in the fish of the coral reef, but is not exhibited by those fish which have an unusual pattern of life or a specially distinctive body shape. Thus, for instance, the pilotfish *(Naucrates ductor),* which attaches itself to sharks and other large fish, does not use countershading. This fish has no need of it, because its position is constantly changing. Sometimes it swims along the host's back, sometimes under its belly, and in consequence its position with relation to the direction of the light is never the same. Deep-sea fish and parasitic creatures are others which do not exhibit countershading.

Imitating the surroundings

Patterning and countershading disrupt the regular shape and symmetrical lines of the body, making the creature's form less conspicuous. In addition to using colour camouflage, many creatures also employ their body surface structure to fit into their surroundings. In many cases the surface of the body is so irregular in shape that it becomes, as it were, a part of the background. But it is not only the surface against which the creature moves which can be imitated by its body form, but many other component parts of its habitat – such as leaves, branches, lichens, algae or the spines of sea-urchins. The value of such camouflage as a means of preserving the species is obvious: the creatures imitate their background, which is of no interest to the predators that feed on them, and this very lack of interest on the part of the predator protects them from being eaten (Plates 75–79). Camouflage, however, is not the monopoly of the creatures preyed upon, since the predators also mask themselves. I once observed in the Red Sea a stonefish *(Synanceja),* which was skulking in a meadow of eel-grass. The body surface of the fish was so remarkably similar to the surface of the growth of algae that with the best will in the world I was unable to determine where the fish's head or tail was. Although I had been observing this meadow of eel-grass daily for a couple of months, each time I had great difficulty in picking the fish out at its chosen station. The stonefish lay absolutely motionless, always at the same place. Once a young wrasse *(Coris caudimacula)* swam over the dangerous predator without seeing it. In a fraction of a second, it had been swallowed in a lightning snapping movement. But the stonefish itself did not once move from its place during the process.

Anglerfishes, too, camouflage themselves from their prey, and are even more cunning in catching their victims. A wormlike process above the jaws presents the prey with the illusion of a titbit. In this way, the anglerfish no longer has to rely on the chance passage of its victims, and instead attracts them by means of its lure.

Off Madagascar I found a shrimp *(Tuleariocaris zanzibarica)* in the forest of spines on a diadem sea-urchin, which it was hardly possible to distinguish from the spines themselves. Its threadlike body resembled the sea-urchin in colour and the shrimp reinforced the camouflage by swimming in a vertical position with head downwards, skilfully spinning around the outer ends of the spines. Any predator out for a meal had no chance of finding this shrimp on its sea-urchin host. But without a sea-urchin as a refuge, the shrimp is defenceless against its foes, as I have been able to confirm by aquarium tests.

At Key Largo (on the Florida coast) I observed a trumpetfish *(Aulostomus)*, which was visibly disturbed by my presence, since it tried to shake me off by a skilful deception. It swam suddenly up to a fan coral and took up between the coral branches a position differing from its normal horizontal swimming position, maintaining instead a vertical station with its head downwards (Plate 79). It then further refined this camouflage, to the extent of even imitating certain of the flexing movements of the coral branches. That it was orientating itself meanwhile was something I established by means of a little experiment. I chased the fish across to a large gorgonian, which possessed several conspicuous 'veins' in its middle section. The trumpetfish sought with all its might to find some possible means of camouflage and finally took up a position parallel to the 'veins'.

Consequently, it is not only the actual physical characteristics which constitute good camouflage, but there can also be specialized forms of behaviour which further refine the visual camouflage.

Camouflage by masking

In the tropical and sub-tropical oceans there lives a group of crabs, known as spider crabs, which cover themselves, by way of camouflage, with all sorts of materials drawn from their surroundings. This group comprises the genera *Hyas, Pisa* and *Maia.* Their principal camouflage material is algae and sponges. If some of these crabs are placed in an aquarium where there is loose material on the bottom, it is not very long before they have once again hidden themselves under their camouflage. With the help of their highly mobile pincers, they arrange the growing camouflage very skilfully on their bodies. To attach these pieces of masking material to their carapace, they have further developed tiny hooks. The *Dromia* crab holds the camouflage material – in this case, generally sponges and bivalves – with its hind pair of legs. Without camouflage material these crabs scratch around unhappily and are obviously ill at ease.

Many creatures are visually adapted to their surroundings without their having deliberately adopted camouflage. Numerous growths – vegetable or animal – become attached to the surface of the body fortuitously over a period of time. Thus the armoured carapace of the large sea-turtle is often densely covered with algae and barnacles without any advantage accruing from this fact.

Generally the creatures which practise camouflage are making use of what are only minor shortcomings in the perceptory system of the creature they deceive. Mutation and selection caused the stimulus-emitting creature to adapt itself in many different ways, in body shape and in behaviour, to these small defects of perception. This is a great achievement of natural selection.

Mating and parental care

The survival of a species in the coral reef – as elsewhere in the sea – is dependent on the number of descendants which a pair can produce. The stock of a given species will not be threatened if at least two of the young from each pair grow to sexual maturity. Fish can produce up to 300 million eggs in a year, and snails as many as 500 million. In spite of these astronomical figures, the numbers of a population in equilibrium with its environment remain approximately constant.

What happens to all these eggs? Why does not each one of them give rise to a fish, a snail, a sea-urchin? The road from the fertilized egg to the sexually mature adult is long and hazardous, and at every turn accidents and enemies can cut it short. This can, in fact, occur even before fertilization.

Mating

When the partners eject sperm and spawn during mating, many eggs will remain unfertilized and will perish. Goatfish, for example, swim in pairs in a steep 'looping-the-loop' movement up to the surface of the water, discharging sperm or spawn as they do so. Only part of the eggs, which float in a whitish cloud of sperm, will normally be fertilized. It only requires the female to eject her eggs a fraction of a second too early during the movement, and thousands of the eggs will perish. Thus the partners must perform in precise synchronization to achieve optimum fertilization. But even when the eggs and sperm are ejected simultaneously, undersea currents can thwart the coming together of the sexual cells.

Many inhabitants of the coral reef mate in this fashion. Every one of them must produce thousands – if not millions – of eggs, to ensure that a just sufficient number is fertilized. But nature has other ways of largely eliminating chance from the fertilization process. Many lipfishes (Labridae) and damselfishes (Pomacentridae) lay their eggs in nests which they have themselves built, or attach them with threads of mucus to the seabed. The male swims many times over the eggs and ejects sperm to fertilize them. Since the eggs are attached, rather than free-floating (as is the case with the goatfish, for example) the majority will be fertilized. The best guarantee of fertilization is internal fertilization within the body of the mother. This type of fertilization occurs particularly with sharks and rays, with sea-turtles and marine mammals. The mating of these creatures under water is a particularly thrilling spectacle.

In the Indian Ocean, I observed the beginning of copulation of a pair of mantas. These huge sea-dwellers 'flew' around each other for a time in elegant curves, then suddenly turned, approaching each other with ventral surfaces facing. In this position they floated motionless for minutes in the water. The underwater photographer Raphael Plante photographed the courtship and copulation of the green turtle *(Chelonia mydas)* in a coral reef off Madagascar. The enamoured pair swam for more than half an hour close under the surface of the water, ignoring the presence of the diver. The male snorted loudly, gripped the female's shell by means of his flippers and seized her throat firmly in his jaws. Meanwhile he introduced his penis into the female's sexual orifice. She left this mating struggle in a somewhat battered condition.

Even when the percentage of eggs fertilized is increased by special mating procedures, this by no means implies that all the eggs will produce adult progeny. Most of the creatures which inhabit the coral reef are the worst type of unnatural parents, by human standards. After mating, they pay no further attention to the eggs which have been committed to the water, and leave them to their fate. Helpless and unprotected, the eggs drift as a part of the plankton. Other species, however, protect and care for their young and in this way increase the number of young raised. The fewer eggs a pair produces, the more attention they lavish on their brood. If the young are left to fend for themselves later, when they have already grown larger and stronger, they have a better chance of survival.

Brood care in invertebrates

Attention and care for the young is by no means restricted to the mammals. Some brittle-stars (Ophiuroidea) are provided with special brooding-pouches on both sides of their arms; in these the eggs are raised to fully-developed young brittle-stars. Whereas in most species fertilization occurs outside the body and many thousands of eggs are expelled, subsequently yielding tiny free-swimming larvae, the number of eggs produced by the species which practise parental care is smaller, and the female expends less metabolic energy on egg production.

Some starfish lay their eggs on the bottom and then crawl onto the egg mass to protect it with their bodies. The octopus links its eggs together like a string of beads and hangs these necklaces in its protected cave, over which it keeps watch. The small octopus species *Octopus aegina* protects its eggs by laying them in a heart cockle (Plate 86). Many deep-sea cuttlefish *(Oegopsida)* immerse their necklaces of eggs in a dense mass of mucilage as a protection against predators. In the Red Sea I once discovered a cylindrical 'sausage' of mucilage a couple of metres in length, floating close below the surface of the water. As I looked at it, with the sun illuminating it from behind, I saw the thousands of shining-red embryos moving along the surface of this cylinder in one direction. A shoal of *Caesio* fish pounced on these opalescent eggs and tried to eat them, but found it quite difficult to penetrate the elastic mass. Whether such masses of eggs, in this case driven by chance to the surface, would, on the deep-sea bottom, be guarded by the cuttlefish has so far escaped observation. Brooding in invertebrates can therefore, as is the case with the brittle-stars, be carried on in specialized body cavities or the eggs can be guarded outside the parental body. In other cases, a suitably safe wrapping provides protection against enemies.

Parental care in fish

The best protection for the brood is the mother's body. In the coral seas there are but few fish which – like the sharks and some species of ray – are known to carry their young like mammals from the egg stage to the birth of the fully-developed creature in the mother's body. The pattern of parental care in the pipefishes and seahorses is most curious. The female inserts her penis-like ovipositor into a slit in the male's belly; this slit is the entry orifice to a brooding pouch, where she leaves the eggs to the care of the male. With these creatures it is the father who bears the burden of giving birth.

Most reef fish give less attention to their young and care for the fertilized eggs only until the larvae emerge. In the process some of them carry their eggs in their mouths. In their jaws their brood is very well protected against being eaten by predators and is in addition continuously provided with oxygen. The cardinalfishes (Apogonidae) and jawfishes (Opistognathidae) are mouth-brooders of this type. In the Indian Ocean I observed a female cardinalfish in the process of laying an enormous mass of eggs. Immediately, many members of the same species threw themselves on the eggs and devoured them. Presumably the female was too slow in taking the eggs into her own mouth and was therefore obliged to leave her brood to her greedy congeners. I have also observed that in the case of cardinalfish of the genus *Rhabdamia,* other members of the species regularly adopted the 'ready-for-feeding' mood when the females were spawning.

The triggerfish *(Balistes fuscus)* lays its eggs in a hole which it builds in the sand, with a diameter of about three feet. For one day the fish watches over and fans the eggs. During this period, it will defend its nesting pit against any invader with the utmost vigour – even human divers swimming by will be attacked (Plate 85). During filming I once ventured too close to a nest and was just able to defend myself against the fish's bites by using my camera. The triggerfish followed me for several metres and positively saw me off.

The damselfishes glue their eggs normally to some firm base, fan fresh water over them at intervals and defend the nest. One species, the anemonefish *(Amphiprion),* lays a particularly attractive, carmine-red egg mass containing some three hundred to four hundred eggs, which it attaches to coral branches under the protection of the sea-anemones (Plate 84). The eggs, which are roughly cylindrical, attain a length of up to 3 millimetres (⅛ inch). The fish watches over its eggs for some ten days. From time to time, it removes with its jaws unfertilized eggs which are dying off, and fans fresh water over the remainder with its fins. Although the anemonefishes drive off all the other fish which would willingly snap them up, the mass of eggs is nevertheless not entirely safe. Small crabs creep up under the shelter of the anemone and frequently manage to plunder the entire mass, a little at a time.

The larvae of the damselfishes (Pomacentridae) generally hatch out at night and immediately drift off to become part of the plankton. At night they are protected from many fishes, which by day feed on the plankton. This is at least a protection against predation in their early hours of existence. But what happens later? Even in the plankton there are many enemies which constantly threaten their lives.

The fate of the young

The fertilized eggs or the hatched larvae are carried along floating with the plankton. It is entirely a matter of chance whether they are transported back to the coral reef by the undersea currents or whether they drift out into the open sea, never to be seen again. While travelling through the water, they feed on smaller plankton creatures – or else are themselves eaten. The free-floating larvae of fish, coral polyps, snails, bivalves, echinoderms or crustaceans play an important part as intermediate links in the interlocking food web of the coral reef, so that ultimately they serve to provide nourishment for their own parents. All this tremendous expenditure of effort on reproduction and parental care was not in vain!

Things are not precisely organized as well as they could be to assure the future of the young, if we consider the course of development from the egg onwards. This makes it comprehensible that, with the existing interplay of chance factors and predation, only the provision of astronomical quantities of eggs will ensure the survival of a species.

Only the marine mammals – more highly evolved – take more trouble in raising their young. Their progeny are suckled and protected. But for this to be so, it has been necessary to develop a highly organized social behaviour providing the progeny with refuge in the group.

87 *(Opposite)* A ribbontail ray *(Dasyatis lymma)* grubbing in the sand for food. The stirred up sand is a signal which attracts other fish, who join in the foraging.

88

89

90

88 The goatfish *(Pseudupeneus macronema)* stirs the sand about, foraging for titbits, throwing out through its gills a cloud of sand which the wrasse *(Coris caudimacula)* picks over for food in its turn.

89 Another species of wrasse *(Cheilio inermis)* swims with a goatfish, accompanying it for long distances, until it eventually drills into the sand and stirs up food for its unbidden guest.

90 Diadem sea-urchins are colonized by many species of cardinalfishes *(Paramia* and *Cheilodipterus),* which find refuge from their enemies between the spines.

91

91 Remoras *(Echeneis naucrates)* have attached themselves to a sea-turtle. This species of fish has a sucker on its head, consisting of a series of lamellae. In East Africa, fishermen use the remora to catch turtles. They attach a line to the remora's tail and throw it into the water when the turtles appear.

△
92 The pilotfish *(Naucrates ductor)* accompanies large free-swimming marine creatures or floating objects, like logs or boats. The photograph shows a swarm of pilotfish with a grouper, at the entrace to its hole. It is not known for certain whether the host fish eat pilotfish from time to time; observations carried out on sharks in the Azores tend to indicate that this is so.

93 The large eagle or spotted ray *(Aetobatus narinari)*, is also accompanied by remoras. They pick skin parasites from the body of their host and enjoy protection in the vicinity of the ray.

◁94 Anemonefishes *(Amphiprion)* live among the stinging tentacles of sea-anemones, which are deadly to other fish. The *Amphiprion* carefully rubs itself against the anemone's tentacles, absorbing in the process a protective substance, which inhibits the stinging cells' discharge. If the anemone had not developed this protective substance, its tentacles would sting themselves.

95 Even at night the anemonefish does not leave the anemone, but hides deep inside its retracted tentacles.

96 The piston-like tips of the tentacles of the sea-anemone are eaten without risk by certain specialized fishes, in spite of the fact that they are poisonous.

95

96

97 Many hermit crabs carry sea-anemones around with them for the whole of their lives, receiving protection from the poisonous tentacles. Some crabs help their anemone partners to climb onto the snail-shell which the crab inhabits – it is even possible to observe signals between the crab and the anemone.

98 A partnership between *Alpheus* shrimps and gobies *(Cryptocentrus sungami)*. The shrimp shovels sand out of the common dwelling-hole, while the fish stands guard outside and gives warning at the approach of enemies.

99 The oral arms of the rhizostome medusa *(Cotylorhiza tuberculata)* are colonized by juvenile horse mackerel *(Trachurus trachurus)*.

97

98

100 *(Opposite)* Cleaning symbiosis. A scad *(Priacanthus hamrur)* being treated by a cleaner wrasse *(Labroides dimidiatus).* The client assumes a position of invitation by inclining its body obliquely.

◁101 The batfish *(Platax)* opens its gill-slits wide to allow the cleaner to enter the gill cavity. Many parasites preferentially settle between the soft plates of the fishes' gills because they find refuge there.

102 A parasitic crustacean (isopod) has attached itself by its sucker to the gill-cover of a whitegrunt *(Haemulon plumieri).* This parasite, which can actually kill the fish, adheres so tightly that it cannot be removed by the cleaner fish. However, the most frequently-occurring parasites are copepods, which run about freely on the body surface.

103 *(Overleaf)* A cleaning station on the reef. A goatfish *(Pseudupeneus dentatus)* stands on its head to invite a butterflyfish *(Chaetodon nigrirostris)* to clean it.

102

4 Partnerships between species

Along the reef, thousands of individuals of different species live together in apparently peaceful coexistence. Certain species learn, for instance, that in the vicinity of another species there is always plenty of food, or they discover that some other species can provide them with protection. Often there are advantages to each partner in forming close symbiotic relationships. On the other hand, there are also cases in which one creature is exploited by another, or suffers from the contact or even perishes. These unpleasant parasitic relationships are more frequent in nature than those in which partnership is of mutual benefit.

The behavioural scientist finds it particularly interesting to investigate the significance and evolution of such interspecific symbiotic relationships, and at the same time to study how the partners communicate.

Firm relationships of this kind between creatures do not come into existence overnight. It takes a long time before the partners become so adapted to each other that their relationship can be transmitted genetically from generation to generation. Casual partnerships can, however, arise within the lifetime of an individual; in this case, learning processes play a large part.

Partnership between fish and man

Let me relate an experience from the Red Sea. For some time I daily fed a triggerfish with sea-urchins. The triggerfish soon learned that the arrival of a large black organism with four spidery limbs, a giant Cyclopean eye and two yellow tubes on its back which constantly spat out air bubbles, signified that something tasty to eat would be offered. Soon it swam patiently behind me, even when I did not give it sea-urchins. A partnership had been formed between fish and man. Finally, it became so importunate, even to the extent of disturbing my experiments, that I had to shut it up in a cage.

This happened three years ago, but even today the same triggerfish swims after every passing diver. When I was there for the last time in October of last year, I immediately recognized my old companion. Its forehead showed the scars of many urchin stings and clearly indicated that it had also begged sea-urchins from other divers. This 'interspecific' partnership began unilaterally with the fish, since it benefited by our association (I will confess that it was not entirely without interest to me, since it was always pleasant to have its company). Our partnership was one which the fish created in the expectation of tasty feeds. Food associations of this kind I have observed in many fish along the coral reef.

Guests at table – commensalism

Goatfishes forage continually in the sand, as I have already described. They take the sand into their mouths, sift it through and finally discharge it through their gills backwards or spit it out onto the sea-bottom again. In doing so they create clouds of sand which contain many small titbits, since they are unable to extract absolutely all the food particles that were present in the sand. These clouds of sand serve to attract many other fish which, in their turn, forage in them. It frequently happens that the goatfish dig so deeply into the sand that they are almost lost to view. What is more, only the guests around their table indicate that somewhere below in the sand is a goatfish digging for its dinner.

The tiny wrasse *Coris caudimacula* is particularly bold and hangs around close to the goatfish. Another lipfish, *Cheilio inermis*, even follows it in open water, swimming in close contact with its body, 'riding' on it in the expectation that it will shortly settle to the bottom and start searching for food again. When the guests become so importunate, the goatfishes drive them off in short order. All the 'dining guests' are fish of restricted range. They frequently wait at the very edge of their territories for the sand-digger to arrive and only leave it once the forager reaches the other edge of their plot. The tiny species of triggerfish *Hemibalistes chrysopterus* is particularly alert and cautious, and while it picks over the clouds of sand for food, it flicks its mobile eyes to left and right, constantly on the look-out for enemies.

Experiments under the ocean

What are the stimuli which set off a unilateral feeding association of this kind? Is it only the cloud of sand which attracts the guests, or can they also recognize the goatfish by particular characteristics? I used a lead weight suspended from a string to produce a large cloud of sand by up-and-down movements, and waited with great interest to see how the fish would behave. In fact, driven by curiosity, a few fish did draw a little nearer, but not very near. Clearly they were aware that the sand cloud, in which they could not see a goatfish, was somehow not quite right. Only when I attached to the lead weight a dummy goatfish with typical body patterning did the customers come close. Apparently, to be stimulated to eat they must have both factors, the presence of the goatfish and its attendant cloud of sand. One triggerfish had, however, detected my deception with the dummy. It considered the dummy and the sand cloud produced by my lead weight to be much larger than ever a goatfish could have stirred up, but contemplated them only from a safe distance and approached no nearer. It would appear certain that for this 'intelligent' fish it was necessary to see specific forms of behaviour characteristic of the goatfish – which I had not been able to imitate with my wooden dummy – to assess the entire situation.

Other feeding associations

Not only the goatfish forage in the sand. Every morning a large grey ray swam across the garden eel area that I was observing in the Red Sea, busily rummaging in the sand for bivalves and small crustaceans. It was always followed by a train of other fishes, which in turn picked over the sand which it had stirred up.

Just recently I found for the first time in the Red Sea another unusual feeding association, this time between two predatory fish. Regularly, late each afternoon, a white-speckled territorial muraena or moray eel *(Lycodontis javanicus)* quitted its accustomed hiding-place to go out looking for food. It ransacked every corner of a large stack of coral, and each time it reached this stack it was joined by two local groupers *(Cephalopholis argus),* which then accompanied it for the remainder of its hunting tour. They swam close to it and waited patiently outside the holes into which it had gone for a moment in search of prey. Now and then the groupers were able to pick up crumbs that the muraena dropped. Since I was later able to observe the same feeding association at other points on the coral reef, I assume that this is not a chance partnership.

Fish that use stalking-horses

Many years ago the famous marine biologist Hans Hass discovered a very curious animal association – indeed a relationship which, just as with the feeding associations, was advantageous to only one of the partners. In the Caribbean, he noted how the trumpetfish *(Aulostomus maculatus)* hid head-downwards between the branches of horn coral. As a parrotfish came by, the trumpetfish left its hiding-place and followed the parrotfish, taking up a position along its back and not allowing itself to be shaken off by anything it could do. I must emphasize that the trumpetfish did not hold on in any way, but simply swam with the parrotfish, close above its dorsal fins. The parrotfish, excited at first, slowly calmed down and the trumpetfish began to follow every movement of its host. Hass assumed that the trumpetfish was in fact using the parrotfish as a stalking-horse, but was at first unable to confirm this.

The trumpetfish went for a ride not only on parrotfishes, but also on groupers or even goatfishes which were much smaller than itself. During subsequent observations, Hass and his fellow-workers noted that the trumpetfish – a voracious predator – is able to creep up unnoticed on smaller fish which it then devours in a flash. The trumpetfish always camouflages itself by hiding behind those species of fish which do not represent a danger to its prey (Plate 11).

In the Indian Ocean and in the Red Sea I have frequently seen isolated young goatfishes on the backs of lipfishes or even of older members of their own species. They swam in close proximity to the dorsal fins, and they too followed every movement and change of direction of the host. Even young of the species, once separated from their group, tried to ride with one another in the same way. Presumably this behaviour comes from a deeply-felt need for protection, as soon as the creatures are separated from the school they belong to. A similar 'desire for company' can be observed in other species which follow large marine animals in the open sea.

Attendant fishes

Pilotfish *(Naucrates ductor)* are frequently seen swimming in front of the jaws of sharks, manta rays or whale sharks. It used to be thought that these pilotfish led their huge companions to their prey. Nowadays we know that pilotfish follow not only other fish, but any fairly large object, such as floating logs, divers or even small boats. I once noticed a shoal of young pilotfish which had crowded together close round the propeller of the outboard motor of our boat (the motor having been switched off). When we caught a few of them and put them in the aquarium, they selected a parrotfish as their companion.

Hass discovered that pilotfish swim into the mouths of large rays and whale sharks in the event of danger, to take refuge. Off the Azores, however, they maintained a position level with the dorsal and caudal fins of certain species of shark and carefully avoided the vicinity of the mouth. No one has yet observed whether pilotfish are eaten by sharks. Their behaviour off the Azores would seem to indicate so. But it is now recognized without doubt that it is primarily protection which the pilotfish enjoy in close proximity to the larger marine animals. At the same time they may be able to pick up scraps of food, which are constantly dropped during the hunt.

Other faithful fellow-travellers of larger marine animals are the shark-suckers and remoras *(Echeneis naucrates* or *Remora remora).* In these fish the forward section of the dorsal fin has developed into a large sucker, with which they tightly attach themselves to sharks and other large animals (Plates 91, 93), and so hitch a ride through the water.

The first time I met these fish was in the Red Sea. While I was swimming along, I suddenly felt something touch my back. Frightened, I put out my hand to touch the place and came into contact with a smooth, slimy body – it was a remora, trying to attach itself to me by its sucker. Again and again it swam towards my chest, a procedure I did not find particularly attractive as I am rather ticklish. For more than half an hour I tried unsuccessfully to shake off my troublesome companion. Fortunately, somewhat later a friend came swimming by and finally released me from the persistent remora.

Hans Hass has recorded that the remora bit the nipples of his diver companion Hirschel, and this subsequently happened to me too. It can be assumed that the remoras are looking for skin parasites on the bodies of their hosts. In addition, they share the host's meals. This I was able to observe in a very impressive manner on one occasion in the Gulf of Mexico. From on board the *Rhincodon* we had thrown pieces of meat into the water for bait to attract sharks. Very soon two small sharks arrived and began to enjoy the feast. They brought with them a crowd of remoras, which immediately left the bodies of their companions and set about busily snapping up small particles of food.

The remoras are not specialized to particular host species. I have seen them on the heads, chest- and belly-plates of sea-turtles, on parrotfishes, on groupers and even on large goatfishes. If they do not find a suitable subject for their attentions, they attach themselves to their congeners. Off Madagascar I photographed two remoras of the same size, one of which had firmly attached itself to the belly of the other! The carrier did its best to rid itself of its passenger, but without success. Its caudal fin was spread wide in a threat posture, but the passenger – firmly attached to the other's belly – was unable to see the signals and threat signs, and so placidly maintained its position.

Behavioural scientists have termed this seeking for a holding-place for the sucker as search or 'appetitive' behaviour. In East Africa, this appetitive behaviour of the remora is exploited as a means of turtle-fishing. The fishermen tie a line around the tail of the fish and throw it into the water as soon as they spot a turtle. The fish attaches itself to the quarry and the fishermen can haul both of them on board.

The remora and the host constitute a transport association, which benefits both partners. The remoras are carried around continually by their hosts, share the scraps from their meals and are moreover exceedingly well protected. Recently it was found that the stomach contents of remoras contained about 70 per cent of parasitic copepods, minute crustaceans which came from the body surfaces of the host creatures. The remoras groom their hosts, and this transport association is consequently an example of true symbiosis.

Other transport associations

Generally it is the larger creatures which provide the 'transport system' for smaller fishes which exploit them. The sea-gooseberry *Coeloplana bannwarthi,* for instance, lives on the spines of the large diadem sea-urchin. At night the sea-gooseberries creep up the spines and from this point of vantage extend their sticky tentacle threads into the water. The dense and mobile forest of spines is wrapped in a veil of small whitish pennants. The sea-urchin is nocturnal and moves out onto open flat areas of the sea-bed to feed. In so doing it involuntarily carries the sea-gooseberries around, and the latter benefit from this movement, since they are continually transported from one good hunting-ground to another.

Large numbers of barnacles – small relatives of the crustaceans, which trap food from the water with finger-like tentacles – live on the shells of the large sea-turtles. For the tiny crustacean this ride is of considerable benefit, since it is constantly being carried into plankton-rich open waters.

A very old and well-tried team, in which both partners benefit, is the association – encountered in many oceans – between hermit crabs and sea-anemones. The sea-anemones attach themselves to the snail-shells inhabited by the hermit crab. Each crab carries the shell around, and the sea-anemone on it. This provides the crab with protection from its enemies, since the stinging tentacles of the sea-anemone keep them at a distance. For its part, the anemone has greater opportunities of finding food as the crab crawls about. The relationship between the hermit crab and the anemone is not always developed to the same degree. In the Mediterranean, a crab will even help its anemone to mount the snail-shell, carefully freeing it from the bottom by feeling and tapping it gently, and the anemone accepts this treatment.

A type of coral living on the sand *(Heteropsammia)* can only continue its existence with the aid of a sipunculid *(Aspidosiphon)* which inhabits its calcareous skeleton. The corals are transported about by the worm and lifted up again, should they happen to fall over on the sea-bed. In addition the worm ensures that the coral does not sink too deeply into the sand.

Sea-urchins as dwelling-places

In the coral reef there is hardly a place that has not been adopted by one or another specialized species as a suitable living-place. Even the thick mantles of spines on the various kinds of sea-urchin are colonized by different organisms. I have already described the *Siphamia* which swim in a tight swarm around the spines of an *Astropyga* sea-urchin, so producing an excellent collective camouflage. This behaviour, however, occurs only in a state of overpopulation, since normally two or three of these small fish carefully position themselves between the spines in absolute safety. But the sea-urchin itself tries to catch and eat them, and if it can reach a fish with its spines, it will stab it and carry it on the spines to its mouth, where it is slowly swallowed tail-first.

The large diadem sea-urchins are by day colonized by numbers of cardinalfishes (Apogonidae), which settle in large groups in the forest of spines on the urchin's body. At night, when the sea-urchin goes foraging in the undersea meadows, the fish leave their prickly refuge.

During the day, the urchin provides an admirable protection from predators and is at the same time a good starting-point for foraging. Whether the sea-urchin draws any benefit from the presence of its residents is still a very controversial matter. Eibl-Eibesfeld has observed *Siphamia* fish cleaning their sea-urchin.

It is not yet possible for us to assign any particular significance to the spiny mantle of the sea-urchin in the over-all ecological pattern of the coral reef. It is not only the cardinalfishes which inhabit the sea-urchins: the young of other species, of crustaceans and even of cuttlefish take up at least temporary positions between the spines. The chances of survival of these creatures would be lessened if they were not able to take cover in the sea-urchin's protective spines. The colonists recognize their prickly host by the dark circular body and by its spines. Consequently, they are attracted by dummies which exhibit the same features. For instance, I set up an experiment in which *Siphamia* fishes were able to choose between two urchin-body dummies, one of which bore only a single spine. Every time the fish swam to the dummy with the single spine.

Defensive-offensive alliances

The provision of protection and cover by the spiny defences of the sea-urchins is not unique, and some specialists have taken even more dangerous places as their living quarters, namely the poisonous, stinging tentacles of medusae and sea-anemones. I describe later (p. 178) the species *Nomeus*, which lives among the tentacles of the *Physalia* or Portuguese man-of-war, these tentacles being capable of fatally poisoning even human beings. The underside of the bell of the rhizostome medusa *Cotylorhiza tuberculata* is also colonized by fishes. How the *Nomeus* manages to protect itself against the deadly poison is not yet precisely known.

We do know more about the defensive-offensive alliances between the sea anemones and the fish *(Amphiprion, Dascyllus, Premnas)* or crabs which live among their tentacles. Investigations by the German marine biologist D. Schlichter have shown what it is that protects the *Amphiprion* fish from the stinging tentacles. The anemones' tentacles would sting themselves, if they had not developed a protective substance which inhibits the discharge of their nettle cells. The tentacle is coated with a thin film of this inhibiting agent. The anemonefish exhibit a special behaviour pattern in which they take some of the substance from the surface of the tentacles and rub it all over their bodies. In this way they prevent the stinging cells from discharging, so that they can move about among the poisonous arms without danger.

To enable me to take certain film shots, I carried out the following experiment in the Red Sea: I cleaned the fishes' skin of inhibitor by means of sand and a piece of cloth. I then anaesthetized the fish and put them back on their own anemone. Within seconds they had been stung by the tentacles, gripped and carried to the anemone's mouth aperture. On several occasions I allowed fish without the protective substance to wake up out of their drugged sleep; they swam straight back to their anemone and were immediately stung. This first of all made them withdraw, and they then tried again, gradually attempting to resume contact with the tentacles. First they touched them with their pectoral fins, then with the underside of the throat and in this way over a period of several hours coated themselves anew with this skin-protecting substance.

It is therefore perfectly possible that other fishes could colonize sea-anemones by behaving in a similar manner, but it would be necessary for them to learn how to obtain the inhibitor for their own use. If they did, they would have to reckon with the fierce resistance of the *Amphiprions,* because these fishes attack anything that approaches the anemone – members of their own species, other fishes of any size, sea-turtles and even human divers. To the accompaniment of loud 'tock-tock' noises, they have bitten me so often that I still bear the scars on my body. They showed a particular tendency to attack their own reflections in the glass of our diving-masks. Ferocious attacks were also made on the silvery, glittering bracelet of my underwater wrist watch.

For more than a hundred years marine biologists have been puzzled by the advantages which the fish and the anemone obtain from their partnership. The anemone offers the fish protection against its enemies. The fish lives between the tentacles and retires inside them at night. It is assumed that, in return, the fish cleans and even feeds the anemone. For, if the fish has come upon a victim out of the anemone's reach, it swims back to the anemone with it, and the latter's tentacles frequently take over the prey. Fish of this species are primarily given to eating small fragments and rarely catch large prey. The feeding has been observed only in the aquarium, and it does not appear to be of significance out on the reef. It is quite out of the question that the fish should lure prey specially on behalf of the anemone. On the contrary, many fishes avoid passing too near the anemones, because they are regularly thrashed by the *Amphiprion* for their temerity.

I have already referred to the fact that the coloration of the fish is treated as a warning in these circumstances. Nevertheless, the fish does confer another significant advantage on the anemone. In the Red Sea we have observed various species (e.g. the butterflyfish *Chaetodon fasciatus*), which showed a preference for eating the tentacles of the anemone. We imprisoned the anemonefishes in a glass jar, and at the very moment when the predatory fish began to tear at the anemone, we released them; they threw themselves upon the anemone's attackers and very quickly chased them away. This made it clear to us why the anemonefish are so aggressive. By driving away the predators who try to eat the anemones, they are also chasing off their own enemies, who are after all destroying their home.

Thus we see that the anemones and the fishes are linked by a firm defensive-offensive treaty, beneficial to both partners.

Collaboration between fish and shrimp

On the sandy or muddy bottom of the Indo-Pacific Ocean there are several species of shrimp which live, either individually or in pairs, with gobies in a common burrow. The fish keeps guard at the entrance to the hole, while the shrimp tirelessly shovels the sand away from the interior. If danger threatens, they both retreat inside.

Magnus carried out detailed investigations in the Red Sea into the life pattern of fish and shrimps, to obtain a picture of the significance of this association. In the course of these studies he discovered a series of astonishing facets of behaviour.

In the Red Sea, *Alpheus* shrimps live with gobies of the *Cryptocentrus, Lotilia* and *Vanderhorsia* genera in domicile relationships (Plate 98). The shrimps' burrows are scattered at random over the sea-bed. Each tunnel contains two shrimps of roughly the same size, and each contains one fish. These burrow homes run under the sand parallel to the surface of the sea-bed and can be up to 70 centimetres (27 inches) long. How do these creatures behave in their subterranean corridors? Magnus observed that the shrimps were sifting over the bottom of their tunnels to find small particles of food. During the daytime they use their claws to remove a quantity of mud from the tunnel, so producing a clearly visible waste-tip in front of the entry. The entrance to the tunnel has to be shifted elsewhere, so that the position of the cavity constantly changes, like a gallery in a mine. Within a single day the position of the burrow entrance can be moved quite a considerable distance.

But what are the gobies doing in the tunnel? The fish lie against the slope of the small waste-heap and take no part whatsoever in the excavation of the tunnel. They snap up small living organisms which keep close to the sea-bed, and pass the whole day watching, keeping their caudal fins as close to the tunnel entrance as possible or even inside it. If danger threatens, they flee head-first into the hole, often badly damaging the entrance in their haste.

The shrimps are generally smaller than the gobies, but they dig burrows large enough for the fish to enter. The latter is constantly caressed by the long antennae of the shrimps and it would seem that the goby enjoys the contact stimuli, apparently not even being disturbed if occasionally one of the shrimps pushes underneath it or thrusts it aside. If the fish is on the watch outside, one of the two shrimps comes to the entrance at frequent intervals and extends its antenna towards the fish's caudal fin, as though wanting to reassure itself that the fish is still there. If the shrimp is at work outside the tunnel, then it maintains constant contact with the fish by means of at least one of its antennae.

The shrimps' unpigmented eyes probably do not see very well. Instead, it can obtain the requisite information about the movement of the fish through contact with its antennae. It can also use them to perceive movements of the water. Magnus considers that the shrimps can in fact interpret the water currents set up by the fins or body of the fish as expressions of warning or calming behaviour on the part of its partner. He writes: 'All the movements typical of the normal long-term position of the fish and its slow movement from one place to another, with particularly gentle rippling of its dorsal and caudal fins, serve as signals to the shrimps that they can without danger go about their business outside the burrow. However, they disappear with lightning rapidity into the tunnel if the fish either darts off after a potential victim or is obliged by the approach of danger to prepare to take refuge in the hole, whether its movements in this case are rapid or slow. In every case it gives advance warning to the shrimps with short, rapid beats of the tail fins. These movements of intention on the part of the fish are a signal to the shrimps to withdraw backwards immediately into the tunnel. They cannot now quit their burrow on their own initiative, without receiving the specific 'all-clear' or calming signal from the fish.

The exchange of such tactile signals enables the half-blind shrimp to move about safely outside its burrow. Without the fish – and its function as an early warning system – the shrimp would be unprotected and at the mercy of its enemies on the sea-bed, which offers it no cover. When it leaves its subterranean living quarters for a short time, it can forage on the surface of the sea-bed, which has an abundance of food to offer. For its part, the fish derives from this partnership the advantage of having a secure sleeping- and hiding-tunnel which it does not have to dig for itself.

Cleaning symbiosis

The removal of parasites and diseased skin tissue from the body surface of fishes is one form of foraging practised by many fish along the reef, several of which have become completely specialized in this function. Cleaning offers advantages for both participants. For the cleaner, the collection of parasites represents a supply of food, whereas the client is relieved of the troublesome organisms which inhabit its skin, and may even bring about its death. It used to be considered that the cleaning activity was nothing more than an insignificant refinement among the curiosities of behaviour evinced by the reef-dwellers. However, when the American zoologist Conrad Limbaugh established that the cleaning shrimp *(Elacatinus oceanops)* can deal with more than three hundred clients in a period of six hours, it slowly became evident how important the cleaners are for the 'hygiene' of many of the inhabitants of the reef.

The skin parasites of fish are principally crustaceans, which live on the surface of their bodies, under the scales or – a particular preference – between the gill-plates. Many parasites attach themselves firmly by suction to the gill-covers (Plate 102) and can cause the death of the fish. The parasites probably produce an itch, since many fish are observed to scratch themselves on hard objects. Some, like the rainbow-runner *(Elagatis)* even make use of the rough skin of sharks as 'sandpaper'. They turn on their sides in front of a shark and rub their backs against it as they swim past.

When Limbaugh removed all the cleaners from two small reefs, within a fortnight all the one-time clients of these cleaners abandoned the spot. Only the home-ranging, territorial species remained. Soon white spots began to be visible on their bodies – the parasites were now able to multiply unchecked. Only some time later did new cleaner fish enter the vicinity of this reef.

More recently the experiment was repeated off Hawaii, with a negative outcome; the 'professional' cleaner of the region is the wrasse *Labroides phthirophagus.* Although all the cleaners on one isolated reef were caught and taken away, none of the clients of that region had left by the end of a month. Because of the isolated position of the reef, there was nowhere else for them to go.

Most of the cleaners are attached to their localities. The clients recognize the spot and come even from far-distant areas of the reef to allow themselves to be cleaned. Frequently there is a queue, and the services of the cleaner are in demand for a considerable time.

In Chapter 2, 'The language of signals', I mentioned that many shrimps, too, are known to function as cleaners. In the Red Sea I observed that the cleaning shrimp living there *(Leandrites cyrtorhynchus)* always lives in close proximity to the professional cleaner fish *(Labroides dimidiatus).* Does the shrimp employ the friendly attracting signals of the cleaner? If so, we can imagine how the shrimps' cleaning symbiosis may have come about – this being a question which is always of considerable interest when we investigate animal partnerships.

One hypothesis would be as follows: the cleaner's clients learn by experience to go to a particular cleaning station, attracted by the signals put out by the cleaner. The customers are in an active state of cleaning readiness and invite the cleaner to go to work on them by assuming stereotyped body postures. In these circumstances they would undoubtedly readily accept the services of the shrimps as well. It is easy for us to imagine that the cleaning invitation signals of the clients are 'parasitically' exploited by the shrimps. The evolutionary process would only need to develop at a later stage signals by which the clients might recognize the shrimps as cleaners. However, if the customers are not in a ready-for-cleaning state, they decidedly do have an appetite for cleaner shrimps, as investigation of their stomach contents has shown.

The development of the symbiosis between the shrimps and their clients could, on this hypothesis, have begun at a point in time when the client fish already understood the amicable signals of the cleaner fishes. I have never observed that the cleaner fishes are in any way hostile to the shrimps – and this is as one might have expected, since after all they both belong to the same union! Undoubtedly, the shrimps and the cleaner fishes do not compete for clients, since there is always an excess of fish ready to submit to cleaning.

104 *(Opposite)* The laterally flattened body of the Moorish Idol *(Zanclus cornutus)* enables the fish to manoeuvre rapidly between the branches of coral and helps it maintain itself on course during swimming, as does the keel of a boat.

105, 106 The spiny pufferfish *(Cyclichthys echinatus)* and the boxfish *(Ostracion tuberculatus)* can manoeuvre very quickly in confined spaces. By means of four 'propellers' – two pectoral fins, one dorsal fin and one anal fin – they can turn on the spot.

107 The grouper *(Epinephelus striatus)*, on the alert in its refuge, falls upon its prey with lightning rapidity. Its massive body and powerful muscles give it rapid acceleration.

105

106

◁108 Adaptation to life on the bottom. Many bottom-dwellers support themselves on the substrate by means of their pectoral fins, or they can even walk on them, using them like legs. The picture of the coral sentry *(Paracirrhites forsteri)* shows how it uses the rays of its pectoral fins as supports, lying motionless for hours on lumps of coral. As with many of the creatures adapted to life on the sea-bottom, the swimbladder – the flotation organ – has become atrophied, since it is no longer needed.

110

△
109, 110 The firefish (Pterois volitans) lives predominantly on the sea-bed, but can also cover short distances swimming freely. Since its swimbladder, too, is atrophied, the pectoral fins have become enlarged like wings to give it buoyancy. It also uses them as nets when hunting small fishes. They hinder it, however, from swimming fast, so to avoid falling prey to its enemies, the firefish has developed a deadly venom.

111 With the poisonous scorpionfish *(Dendrochirus)*, too, part of the pectoral fin has become extended to form a supporting surface which it can use to hold itself above the bottom.

111

112 Poisonous creatures. The scyphozoan *(Dactylometra quinquecirrha)* of the tropical oceans fishes a large volume of water with its tentacles. The poisonous tentacles trap plankton and even relatively large fishes.

◁113 The cone-shell *(Conus textile)* injects its victims with a poison from its salivary glands, as though using a hypodermic needle.

114 Along the Australian reef lives the poisonous olive sea-snake *(Aipysurus laevis)*. Sea-snakes have even attacked divers in that region.

115 *(Overleaf)* The dangerous tentacles of the Portuguese man-of-war *(Physalia physalis)* stretch for as much as 50 metres (160 feet). Their sting can be fatal to human beings.

5 Adapting to the environment

During my various passages through the reef I often asked myself, what is the purpose of the multiplicity of forms which these creatures exhibit, what end do all these strange body shapes serve, and for what reason has nature produced such strange figures as the seahorse, the firefish or the boxfish? Is it possible for such fish even to maintain themselves in the struggle for existence in the reef? Is their design valid? It is an undoubted fact that given habitats on the reef bring into being quite clearly-defined body shapes. The familiar 'normal' fish form, resembling that of a herring, is primarily characteristic of the open water. Most of the variant types – the flattened, high-backed, globose, spindle-shaped or box-form creatures – are, on the contrary, found everywhere throughout the jungle of coral branches. Over the sandy bottom, again, the typical fishes have depressed, low-profile forms similar in shape to a pancake, a form which constitutes a special adaptation to this particular habitat. What pressures of selection have been active to produce this multiplicity of forms?

The body and fin forms of a fish – and here we are speaking only of fish – can be the consequence of a very varied range of factors: they may be imposed by the techniques of food gathering, may result from the particularities of the habitat occupied, and may also be a consequence of the social partners with which the fish is associated. The small parasitic fish *Carapus,* for instance, lives in the rectum of sea-cucumbers. It leaves the body of its host only at night, as I have on many occasions observed. The rear part of its body has atrophied in the course of its development to its manner of life and no longer possesses fins, which would only be a hindrance in its narrow home. In many cases, it is easy to see how special forms of body and fin have been developed to suit the type of movement employed by a particular species. Water is eight hundred times as dense as air; in order to overcome the resistance which it offers, the fish require powerful propulsive organs. The main driving power is generally provided by the caudal fin; lateral beating movements produce thrust forces which propel the body forward. In many fishes, however, the caudal fin is not fully developed, and such creatures progress by undulatory movements of the rear of their bodies, while others row themselves through the water with their pectoral fins.

High-speed swimmers of the free water

In the free water above the reef there are no hindrances such as exist in the confines of the reef. The fish can move freely and do not need to turn sharply as they hunt their prey. In these wide open regions they can only catch their victims by moving at high speed. Certain predators achieve speeds of up to 80 km. an hour (50 m.p.h.), a velocity which is possible only with a slender, streamlined body. The main driving power comes from the upward-pointing caudal fin and the muscular root of the tail.

Species of shark and mackerel belong to the high-speed swimmers of the reef. Their pectoral fins are the control surfaces, which act like the ailerons of an aircraft. One disadvantage of this design, however, is that the fish can make only minor directional changes by changing the position of their fins.

Wanderers of the reef

Many damselfishes and snappers wander in large groups through the reef or swim directly along the edge of the reef, searching for food. At the approach of an enemy they very rapidly hide in the fissures in the reef. Their manner of life requires not only that they should be long-distance swimmers in order to reach new sources of nourishment, but also that they should be capable of making a standing start, to escape from their enemies. These fish are highly manoeuvrable, and their equipment of fins enables them to swim rapidly and to accelerate quickly.

Masters of manoeuvre

In the thickets of coral branches there live the many high-backed species of fish, such as the brightly-coloured butterflyfishes or the Moorish idol (*Zanclus cornutus,* Plate 104). The surgeonfishes, which are specialized for grazing on the algae and micro-organisms found in this highly structured region are flattened like a discus. This body form enables them to turn rapidly on the spot and at the same time assists them to maintain a steady course, in much the same way as a boat is steadied by its keel.

These fish are not long-distance swimmers. They have large pectoral fins, with which they can row, and large dorsal and anal fins, which provide the necessary drive. In flight, many of them progress by undulatory movements, that is, they behave like fishes of 'normal' shape. Additional aids towards a greater degree of manoeuvrability in confined spaces have been provided by nature for the boxfish and globefish. These box-shaped or spherical fishes have four propellers – two pectoral fins, one dorsal fin and one anal fin – with which they can even turn on the spot. It is always an amusing sight to see these fish in flight. They make great efforts to accelerate, making powerful strokes with their rounded, supple tail-root, but even this does not enable them to move particularly quickly.

A specialist in movement of a particular kind is the sea-horse. Its dorsal, pectoral and anal fins are arranged in different directions, so that it can move in any direction without changing the axis of its body. At the same time it moves buoyantly towards its prey, which it swallows with its tube-like snout. The sea-horse is unable to make rapid forward movements because – like the boxfishes and globefishes – it does not possess a powerful drive. This also means that it is unable to flee rapidly. But nature has provided it with protection of another kind, since the sea-horse camouflages itself from its enemies by the colour and texture of its indigestible, armoured skin.

Quick-start artists

Groupers (Plate 107) and scad (*Priacanthus hamrur,* Plate 63) lie in ambush for their victims, taking them by surprise with a rapid dash. Many other bottom-dwelling predators have developed similar hunting methods. The quick start they need for the purpose is achieved by using all their available fins, and in addition they beat powerfully with the entire rear portion of their bodies. They are decidedly 'sprinters' and as such are unable to sustain this speed for any length of time. If they have to cover long distances, they repeat their acceleration movement and then glide on without moving their bodies. Since this type of lightning start, with all fins working, calls for a powerful muscular system, these fish are generally compact and powerful in build.

Tube- and hole-dwellers

I am repeatedly surprised by the playful ease with which moray eels can swim through the narrowest fissures and tunnels. Only a body built like a snake's could possibly negotiate such narrow spaces. The muraenas move by means of long narrow fringes of fin along the body, but also make use of the many obstacles in their habitat to help them to advance. In doing so they can change their direction of movement at will and can even creep backwards. Nature has seen to it that all the projecting fin structures have been selected out in these creatures, because they would only be a nuisance in the confined world which they inhabit.

Even these few examples will have shown clearly how the body shape and fin form of fishes have arisen in the course of adaptation to their manner of life and of progression. The fins are, moreover, not exclusively organs of locomotion, and can in fact perform other functions. For example, they can serve as signals to attract or impress their partners in the mating season, and as a threat signal in combat. Sometimes, too, they are transformed into suckers, as with the suckerfish or remora (*Echeneis,* Plates 91, 93).

The shape of the firefish

One fish which lives in the coral reefs even uses its fins as a means of catching its food. In its effort to adapt to its special mode of life, this fish has developed a shape so bizarre that it has kept hardly anything of the classical fish appearance. This is the well-known firefish or long-spined turkeyfish (*Pterois,* Plates 109, 110), whose ragged-edged wing-like pectoral fins make it look like a moving, magnificently-coloured but highly dangerous bush. Konrad Lorenz has carried out investigations into its behaviour and development.

The firefish belongs to the poisonous Scorpaenidae or scorpionfishes, whose members have mostly gone over to bottom-living habits, which has led to atrophy of their swim-bladder, the flotation and equilibrium mechanism of free-swimming fishes. Scorpionfishes sit with the hard rays of their pectoral fins supporting them on the ground and patiently wait for a small fish to swim unsuspectingly past them. Since they are very bad swimmers, they have adopted an efficient camouflage to help them in hunting their prey.

Among this group of cunning predators which lie stationary in ambush on the sea-bottom, only the firefish has resumed the free-swimming habit. It has maintained the specialized features developed for bottom-living purposes, including the much reduced swimbladder. In order to be able, despite this, to float in the open water, it has converted its pectoral fins into buoyancy surfaces; these, however, at the same time rob it of the most important faculty of a free-swimming fish, namely rapid progression. The firefish is unable to pursue its prey, and would even be at the mercy of its enemies, had nature not conferred upon it strong poison spines, with which it is able to defend itself most efficiently in the event of danger.

But how can this fish be a predator and catch prey at all, when it is one of the slowest-moving fish on the coral reef? Konrad Lorenz has written on this subject: 'This they do in a unique manner. They extend their front extremities wide, stretching them out rigid at right angles to the body and thus drive their victims into a corner in exactly the same way as one of our zoo assistants might shoo a goose or a crane into the desired place, namely by extending the upper extremities and making compensatory movements to prevent every attempt made by the driven animal to escape to right or left of its pursuer. In doing so the fish – once again exactly like a human being driving an animal – do not advance as fast as some predatory animal which seeks to run down its prey in an open race, nor do they move as slowly as a predator which seeks to creep up on its victim unobserved.' For this purpose the firefish use the pectoral fins as a driving net. The cornered fish is presented with what looks like a way of escaping from his predicament – and this turns out to be a trap. The fact is that in the vicinity of the firefish's jaws the pectoral fins, made wider by flaps of skin, contain a transparent window towards which the trapped fish flees – in the hope of escape. But in fact he is swimming right into the firefish's mouth.

Competitive adaptation between hunters and hunted

Every creature on the coral reef must each day fight for its existence, in a genuine life-and-death struggle. Even predators can become victims of other predators. Each living organism is faced with the same problem: how to obtain its daily nourishment and at the same time to evade the attacks of its enemies, which threaten its very life. I personally am less interested in the manner in which a predator traps its prey than in the protective adaptations assumed by the hunted. Predator and prey are engaged in a continually inventive competition to find new ways of adaptation, in the course of which the potential victims must always be one step ahead of the predator to be able to survive. For instance, where flight is pointless because the predator can swim more rapidly, protective adaptations must be developed.

One of the most vital discoveries to deceive the enemy is that of camouflage. But other methods have proved no less successful. In the event of danger, crabs slip into their holes or rapidly dig themselves in. The hermit crab withdraws into his snail-shell home, and the giant clam or thorny oyster snap their valves closed at the slightest disturbance.

Most reef fish evade danger by retiring to hiding-places which they know with great accuracy, since they have an excellent locational sense. Garden eels disappear backwards into their tubes in the sand. Soles and flounders wriggle into the sand, while sand-eels dive headfirst into the sandy bottom. Even schooling behaviour (p. 201) is an adaptation in the face of the enemy.

The invertebrate and its enemies

The bottom-dwelling invertebrates are threatened by particular dangers from enemies that feed on them – mostly fish – because they do not possess the facility for rapid flight. Protective adaptations are therefore particularly noticeable in these cases. I have already described how the poisonous tentacles of the sea-anemones are consumed by certain specialists. The sea-anemone has therefore developed the capacity for protecting itself against the attack of such enemies by rapidly retracting its body. The coral polyp, too, disappears into its small, calcareous refuge when it is touched. Sponges possess bitter, pointed skeletal elements which are embedded in the interior of their bodies, so that the mealtime enjoyment of those who eat them will be spoiled. But it is not only fishes which are enemies to the invertebrates, since even the invertebrates themselves prey upon other invertebrates. Thus for instance some starfishes find bivalves a delicacy. The scallop *Astropecten* has learned to detect its enemy by scent. As it flees, it flaps the valves of its shell jerkily and hops away from its enemy over the bottom. The predatory harlequin shrimp *(Hymenocera)* in the Indo-Pacific region has become specialized for eating starfish (Plates 143–146). It detects its prey at a distance with the feeler-like antennae, creeps up to it along the scent trail and braces itself against the bottom with its great claws, holding the starfish and tipping it over meanwhile with its legs. Then with its other, tweezer-like claws, it eats the starfish – first the tiny white tube-feet and then its viscera – the gonads are a particular delicacy.

In the aquarium this shrimp has even managed to make a meal of the poisonous spiny starfish *Acanthaster,* the enemy of the coral polyps. Unfortunately it has not yet proved possible to demonstrate with any certainty whether the shrimp behaves in the same way in the wild. If this were so, it could make a decisive contribution to rescuing the coral reefs from the threat posed by the spiny starfish.

The end of a sea-urchin

Many examples of this kind could be given. I will only quote one more, which shows an astonishing degree of behavioural specialization. In the Indian Ocean and also along the Red Sea I repeatedly found the remains of the spines of the great diadem sea-urchin scattered about the sea-bottom. Had these sea-urchins died a natural death or had they fallen victim to enemies? The answer to this riddle came by chance. I was observing a large blue triggerfish *(Balistes fuscus),* which was stubbornly working away at a fissure in the rock where a sea-urchin had hidden. First of all, the triggerfish nibbled away the brittle, pointed spines of the urchin one by one. Then, very carefully, it lodged its teeth in the firm stump of one of the spines and used this hold to draw the urchin out of its hiding-place. The urchin fell to the ground and ran on its spines towards the safety of the rock. However, the triggerfish did not let its prey escape; several times it blew a powerful jet of water at the urchin with its mouth (Plates 140–142). The sea-urchin was not prepared for this sudden strong current and was bowled over. In a twinkling the fish pounced on the urchin and bit its victim on the oral surface, where its spines are shorter. A few moments later the sea-urchin's shell was pierced by the triggerfish's powerful jaws, and it set about ravenously devouring the tasty viscera.

Other ways of hunting sea-urchins

The small triggerfish *(Balistapus undulatus),* which has a much smaller mouth, is quite unable to produce a jet of water powerful enough to knock the sea-urchin over. But I once witnessed the way in which this species can achieve its ends. It nibbled off the urchin's spines, bit into the stump of one spine and carried the sea-urchin out of its refuge into the open water. There it let its prey fall. While the sea-urchin was slowly sinking towards the bottom, the triggerfish swam past it to the bottom and took a carefully directed bite at the vulnerable oral surface of the urchin as the victim floated down. The meal was completed only a few seconds later.

On one occasion I set out some sea-urchins on the open sandy bottom. Quite soon a lipfish of the genus *Cheilinus* arrived and began to swim round the constantly moving bundle of spines. The spines followed every movement of the fish with defensive movements. Suddenly the lipfish laid itself on its side on the bottom and overturned the sea-urchin with a flick of its head. In a trice it bit the urchin on the oral surface and, holding its prickly burden in its jaws, made off with it. I followed the fish's course to a smooth block of coral and there I could hardly believe my eyes, as it began to break the forest of spines into tiny pieces by sideways jerks of its head, knocking the urchin against the coral block. I watched open-mouthed as it then swallowed the various parts of the sea-urchin, together with the remains of the spines.

Subsequent observation showed me that some of these creatures broke up the sea-urchins they caught in this way only at quite clearly defined places in their territory. Thus they made use of some object as a tool to prepare their victim for eating.

Protective mechanisms of the sea-urchin

To the diver who visits the coral reef it is a familiar picture to see sea-urchins congregating in compact groups during the daytime. They interlock their spines, to make it impossible for a predator to carry off members of their group as a meal. If a sea-urchin is on its own, it hides and avoids the open spaces of the bottom. It also avoids, if at all possible, places where there are strong currents. I have already mentioned that the sea-urchin tries to ward off the attacker by defensive movements directed towards the enemy. When do the sea-urchins find time to eat for themselves seeing that they live in this particular manner specifically developed as a protective adaptation against their enemies? They feed on the surface of the sea-bottom; since, however, it is impossible for them to obtain enough food during the day because of the constant dangers which threaten them, they have become nocturnal. After sundown, when their enemies are asleep, they creep out in crowds to forage; in this way they evade the daytime predators.

I have often wondered why it is just the night time which is the great period of activity for the invertebrates along the coral reef. Where, during the day, there is no sign of life, suddenly sea-cucumbers creep out of the sand, sea-anemones extend their tentacles around them into the water and many other echinoderms go on their nocturnal travels. It is certain that many of them – like the sea-urchins – have been forced, by pressure from their enemies, to adopt a nocturnal life pattern on the reef. In this way the periods of activity of the predators and their prey have become staggered in time – an effective method of allowing the hunted to escape the hunter.

116 *(Opposite)* Plankton trappers. The beautiful crown of tentacles of the tube-worms is a plankton trap of very sophisticated design.

117 The sea-pens (Pennatularia) fish plankton out of the water currents with their polyps.

118 Sponges have many forms. They suck in water through their pores and filter out the plankton by means of flagellae.

119 The vestlet-anemone *(Cerianthus membranaceus)*, like the coral polyp, catches its plankton prey with the nettle cells on its tentacles.

120 Tube-worms make close-meshed filter barriers with their crowns of tentacles.

121 Brittle-stars (Ophiuroidea) use their long arms to filter the plankton out of the water currents.

117

118

119

120

△
122 An as yet unidentified creature (probably a specialized anemone) uses its long tentacles to fish for plankton during the night. The plankton is trapped on the tentacles, the arm itself retracts and carries the food to the mouth.

123 The feather-stars (Comatulida) extend, during the night, filtering fans which trap the current-borne plankton. By day these creatures remain in their hiding places with tentacles rolled tight.

124 The feather-star *(Heterometra savigny)* climbs to elevated spots – in this case a brightly-coloured gorgonian *(Lophogorgia)* – to spread its arms out as a plankton filter. The gorgonian itself also orientates itself in the direction of current flow, to fish for plankton.

123

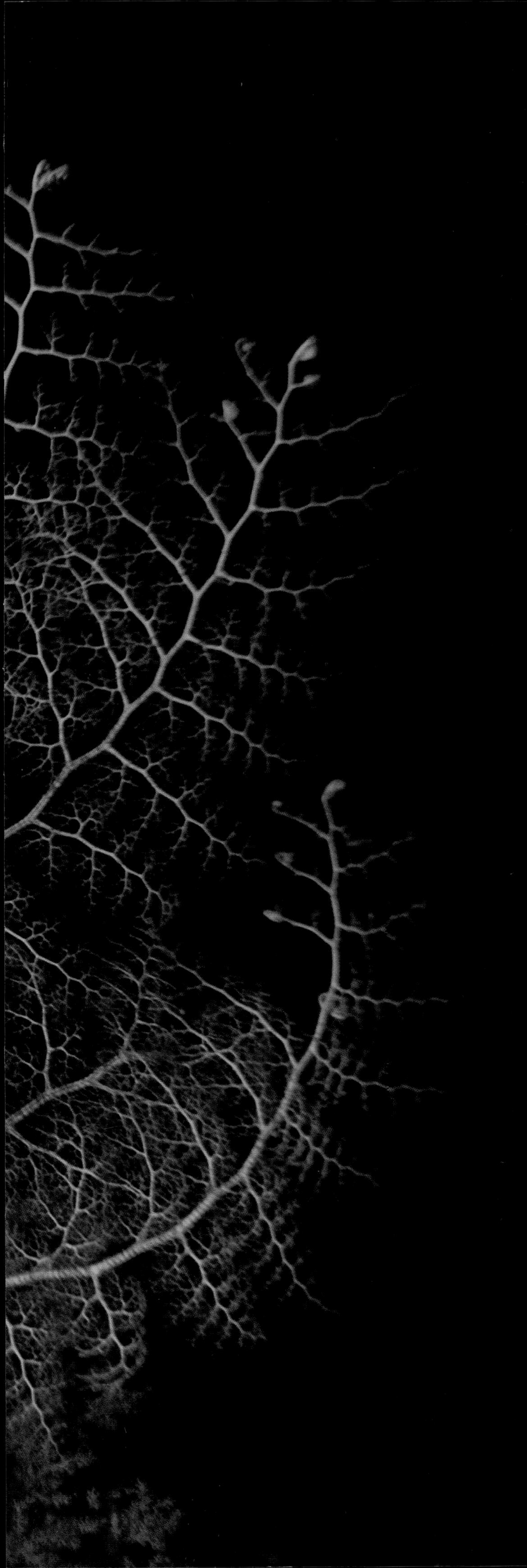

126

127

◁125 The gorgon's head *(Astroboa nuda)* from the Red Sea is a specialized brittle-star (Ophiuroid), adapted to the nocturnal trapping of plankton. Its arms ramify many times to form a giant filter screen, which can be 150 cm. (5 feet) across. The ends of the arms are provided with tiny hooks, which seize the plankton.

126 If the beam of an underwater lamp falls on a gorgon's head busily engaged in filtering, the arms retract spirally. The creature then begins to move across the reef like a walking bush to escape from the light.

127 By day the gorgon's head – in this case *Astrophyton muricatum* – remain rolled up in their rocky hiding places or perch on corals. They do not leave their selected site.

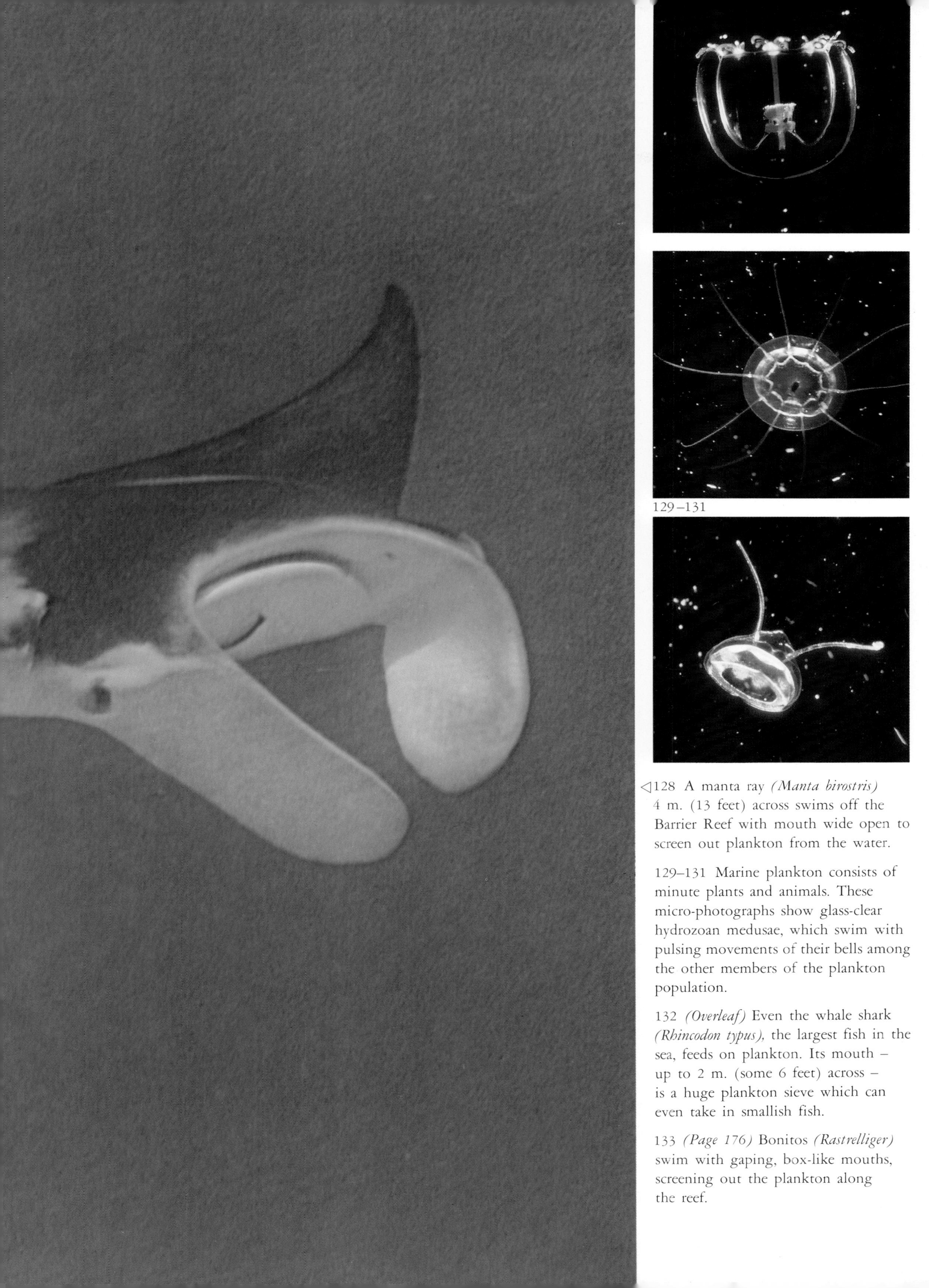

129–131

◁128 A manta ray *(Manta birostris)* 4 m. (13 feet) across swims off the Barrier Reef with mouth wide open to screen out plankton from the water.

129–131 Marine plankton consists of minute plants and animals. These micro-photographs show glass-clear hydrozoan medusae, which swim with pulsing movements of their bells among the other members of the plankton population.

132 *(Overleaf)* Even the whale shark *(Rhincodon typus),* the largest fish in the sea, feeds on plankton. Its mouth – up to 2 m. (some 6 feet) across – is a huge plankton sieve which can even take in smallish fish.

133 *(Page 176)* Bonitos *(Rastrelliger)* swim with gaping, box-like mouths, screening out the plankton along the reef.

6 Plankton traps

Along the coral reef the expenditure of plant and animal life is on a prodigal scale. What then is the purpose of all this expenditure, when each living organism is doomed simply to be eaten by another? Plants and animals are interwoven in the eternal cyclical exchange of energy and mineral substances; in a word, they are elements in the food pyramid.

The broad base of the food chain is formed by the masses of microscopically small creatures and plants, the plankton organisms – hardly visible with the naked eye – which can be present in a single drop of sea-water in thousands and, as the ladder of interlinked food chains progresses, are eaten by organisms somewhat larger than themselves. If we move up towards the apex of the pyramid the number of consumers continues to fall off, while the body size of the creatures increases. At the other end of the chain are the giants of the sea, whose existence would be quite impossible without the presence of all the smaller species.

In a tropical ocean the life of the animal world – just as on land – would be inconceivable without plants, since these are the only organisms capable of converting solar energy to sugar, starch and protein. The marine plants, however, must not be imagined to be like land plants, with roots, branches, flowers and leaves. The majority of marine vegetation consists of speck-sized single-celled organisms, the algae. Diatoms form by far the greater part of the whole. They are surrounded by vitreous capsules consisting of mineral matter precipitated from the sea-water. Nature has given the fullest rein to fantasy in the development of the – frequently grotesque – forms of these casings. The algae form the basic nourishment for the animal plankton carried along in the sea with them, and consisting of an inconceivable variety of living organisms, representing almost every phylum of the animal kingdom. Minute medusae, sponge larvae, bryozoans, hydrozoans, larvae of coral, polyps of echinoderms, of snails, of bivalves, of polychaetes (bristleworms), and tiny predatory arrow-worms which hunt crustaceans – all these form part of the plankton 'soup', and finally there are the fish larvae, which I have already discussed. The copepods are particularly numerous in the plankton and are of considerable importance as nourishment for many reef-dwellers. The existence of all these planktonic creatures would be threatened if the algae which float with them were absent.

There are almost as many kinds of trap developed by the larger creatures to catch the plankton as there are forms of plankton itself – and they are very numerous. Even the 'plankton corpses', the detritus which rains down onto the sea-bottom, are eaten. The plankton-trappers have developed a very varied range of methods and have adapted themselves in behaviour and body structure to their particular pattern of feeding. Frequently they no longer need to move, but pick out of the undersea currents the plankton carried along by them or trap the particles slowly sinking towards the bottom. Many of these creatures therefore sit tightly attached to the sea-bed and so closely resemble plants that one can hardly imagine that they are in fact animals.

Poisonous tentacles

One plankton trap is found everywhere on the reef: stinging, poisonous tentacles which shoot tiny 'harpoons' from specialized cells, and hold the plankton tightly. This multi-purpose device can also be used for defensive purposes. Stinging tentacles have already come to our notice in connection with the coral polyps. They are also used by sea-anemones (Plates 119, 122), and medusae. This type of plankton-catching system is a monopoly of the coelenterates, or cnidarians (the jellyfish, corals and anemones).

The tentacles of the medusae are particularly impressive, and include huge curtain-like hanging processes which trail freely in the water (Plate 112). Contact with these stinging arms can be dangerous even for human divers. The venom of the medusae of the *Chironex* and *Chiropsalmus* genera can in some circumstances be fatal. The poison of the sea-wasp, a medusa found along the Barrier Reef, and which is only an inch or two across, is particularly feared. It is reported that contact with this jellyfish has caused death within periods varying from thirty seconds to three hours. Human beings are also said to have been killed by coming into contact with the tentacles of the Portuguese man-of-war, *Physalia physalis* (Plate 115). Only once did I touch one of these creatures, out of pure carelessness, with my right hand; a burning pain shot through my hand and immediately the points of contact between the tentacles and my skin became visible as red stripes. It was only some hours later that the pain began to abate.

The Portuguese man-of-war consists of a large number of social polyps exhibiting division of labour. The tentacles, which can be up to 50 metres (160 feet) in length, belong to specialized polyps which occupy themselves only with trapping food. The tentacles are retracted spirally at regular intervals, forming small clumps, so making the prey accessible to the polyps specialized in digestion and assimilation of the food. Despite the highly poisonous nature of the tentacles, one species of fish uses them as a refuge. This is the man-of-war fish *Nomeus,* which is apparently unaffected by the poison. A water turtle will even eat an entire Portuguese man-of-war, tentacles and all. An even stranger phenomenon has been recorded: young octopuses of the species *Tremoctopus violaceus* collect portions of the tentacles and use them as weapons of defence or offence.

Ciliary feeders

The sponges which are found on the reef exhibit a multiplicity of forms (Plates 42, 118). They produce magnificently coloured carpets across dead corals or stand clear of the ground in striking trumpet-forms. Some are of fine filigree structure, like coral. If we observe all these varying structures more closely, we see tiny pores, which cover the entire surface. These sponge animals suck in water through the pores. The interior of their bodies is divided into many small chambers, the walls of which are clothed with flagellae, which maintain the water in the cavities in movement and screen out the particles of food. In this way the sponges uninterruptedly draw in eddy currents of water containing nourishment and eject the filtered water. In shallow bays, the surface of the water can seethe under the whirlpool-creating action of the sponges.

Other organisms, like the tube-living worms, stretch out around them beautifully coloured crowns of tentacles, bearing a large number of cilia (Plates 116, 120). The tentacles form a funnel leading to the creature's mouth; the cilia feed a whirling current of water into the funnel, and special grooves carry the sieved-out food into the mouth.

Certain species rotate their ring of tentacles slowly; this graceful movement has often led me to swim close to these creatures, but I have only rarely managed to inspect the crowns of tentacles at close range. If a shadow suddenly falls on them or unexpected vibrations occur in their vicinity they retract rapidly; all that remains is an inconspicuous tube, which has given them their other name of tube-worms. Whereas these creatures themselves stir the water in the process of catching their food, there are others which simply extend their filtering screens or cones into the natural currents in the water and passively trap the plankton they find there.

Current-screeners

Off the coral reef of the small island of Nosy Iranja near Madagascar I found on the sandy surface thousands of arms 5 to 7 centimetres (2 to 2½ inches) long, projecting stiffly upwards. As I sought to grab one of these arms, it disappeared with extreme rapidity into an underground tube. These arms – which can reach a length of as much as 40 cm. (15 in.) – belong to the then unknown species of brittle-star of the genus *Amphioplus*. Using only the tips of the arms, this creature was filtering the stirred-up detritus which had been whirled off the sea-bottom by the currents. Minute tube-feet stick out of its tentacles, set close together and forming a filtering comb. If a particle of food strikes one of them, it is passed from one foot to another down towards the tube where it slowly disappears.

This system of obtaining food by filtering it out had already been observed in other species of brittle-star. One group has many sub-divided arms, and resembles a delicate fern (Plates 125–127). These creatures are known as gorgon's heads after the snake-headed monster of Greek mythology.

When I first met a gorgon's head at night I had no idea that it was an animal. Only when my underwater lamp completely illuminated the many-branched arms did a sudden movement occur (Plate 126). The various branches of the arms rolled together in spirals, and the entire creature crawled across the reef with an undulating movement, like a living bush. This observation fascinated me so much that I decided to investigate its manner of life more closely.

By day the gorgon's heads sit in dark hiding-places and only come out, with clockwork precision, at sunset (Plate 127). They seek out an elevated site and there unfold their arms to form a beautiful filter-fan. A mass of arms forms an almost circular canopy, looking like a large radar dish aerial. The aperture of this canopy is always pointing into the current. Any particles of plankton which make contact are seized by small pointed hooks until the arm rolls itself up and bends towards the mouth, where the food is removed. Shortly before sunrise the entire filtering canopy folds together and the gorgon's head creeps back into its refuge.

To enable me to distinguish between the different individuals, I tried to mark some of them with strips of adhesive plaster, but since the creatures themselves removed these strips continually, I resorted to amputating short pieces of arm to enable me to identify them. This made it possible for me to observe that the gorgon's heads remain at a fixed point and keep to the same home, as well as to the same elevated filtering station, for months at a stretch. I was unable at the time to discover what mechanism they employ for their astonishing homing capacity. The brightly-coloured sea-lilies (crinoids) also creep out of their hiding-holes at night and they too stretch out their arms as filtering fans (Plates 123, 124). Often they mass close together, so forming several large filter barriers.

Magnus has carried out laboriously precise investigations into the food-catching methods of these creatures. The particles of food remain suspended from the arms of the crinoids and are carried by ciliated grooves to the mouth. Presumably not all the particles of plankton benefit the crinoid. Fish of the species *Lepadichthys lineatus* specialize in eating parts of the arms of the crinoid and possibly therefore steal the food trapped there in addition. During the night, we can often observe many passive current-filtering creatures, like the sea-pens (Plate 117), which put out their filtering 'leaves' in the current. Presumably their nocturnal life constitutes an adaptation to the considerably greater abundance of plankton found above the reef at night.

Sticky traps for plankton

The nocturnal plankton-trapping organisms do not only employ filtering canopies or cones for the purpose; some of them have a quite exceptional and particularly cunning method of catching their prey. During a night dive in the Gulf of Aqaba in the Red Sea, I once discovered fine white-plumed tentacles extending from a coral *(Sarcophyton)*. These tentacles reached out in a continuous movement some 1½ m (5 feet) into the water and then retracted with a rhythmical motion. The gleam of my lamp attracted thousands of plankton particles, which formed a dense cloud in front of me and ultimately adhered to the tentacles. It was only some days later that I solved the riddle. I had discovered a hitherto unknown species of ctenophore *(Coeloplana)*, which lives under the canopy of the soft *Sarcophyton* coral. Subsequently I found on the Banc d'Entrée off the Madagascar coast three other species which also go plankton-catching by night. The tentacles are unrolled from a narrow pouch and drawn out into position by the current; they bear adhesive cells which grip any plankton which touches them – much as a fly adheres to a fly-paper.

At one point on the reef the ctenophores had crowded close together, so that their tentacles formed a close curtain. Among them I observed a fish some 10 cm. (4 in.) in length, which I had frightened with my lamp. The more it struggled, the more it became enmeshed in the glutinous tentacles. But all the efforts of the jellyfish were in vain, since it was unable to do anything with so large a prey, being adapted only to handle microscopically small victims. They wipe off the plankton which has been caught on the tentacle pouch and from here it is carried to the mouth by movements of the cilia.

The free-swimming relatives of *Coeloplana,* the true comb-jellies (ctenophores), are glass-clear delicate organisms which move through the water propelled by pulses of their ciliated combs. They swim with the plankton and fish their prey – copepods, arrow-worms, sea-butterflies and numbers of larvae and other marine creatures – from the water around them. The victims, adhering to the tentacles, are later licked off at the mouth.

Plankton-swallowers

Larger plankton-feeders, particularly the copepods, are delicacies for many fish which hunt them in the water above the reef. If these fish suddenly shoot steeply upwards, making a snapping movement of their jaws, this is an indication that they have just had a bite. Many damselfish hang in dense swarms over the reef. In their hunt for plankton they move in a spell-binding dance, their shimmering, brightly-coloured bodies dipping and rising playfully. But this picture of innocent busyness entirely masks the fact that the fishes must daily go hunting uninterruptedly to fill their bellies with their minute prey.

Other plankton-fishers have already come to our notice as denizens of the sandy areas – the colonial garden eels (Plate 44). These occur only on slopes of sand or mud which lie athwart the current. For a total of over two hundred hours I have observed the species *Gorgasia sillneri* in the Red Sea, in order to study more closely these semi-sessile creatures whose manner of life is quite unique among the vertebrates. The thin, serpent-like bodies wave gently in the current, while the eels' heads nod slightly as they snap at their prey. Once thousands of pteropod snails swam past my vantage point. The eels snapped eagerly at such tasty prey, carefully aiming at each individual mollusc before snapping at it. If they missed and their intended victim escaped them, they reached out of their tubes and followed the small snails as far as possible, but without entirely leaving their homes. Frequently only some 10 centimetres of the tail portion of an eel almost a metre long (about four inches in a yard) remained in the tube. Meanwhile the eels gripped the inner walls of their tubes by means of their erected dorsal fins.

The garden eels have fixed feeding times which vary with current movements and availability of food. Shortly before sunrise they put their heads out of their underground homes and start feeding intensively. By late morning their first meal is ended. They disappear again below the sand, and then come out again late in the afternoon. During the full moon, they are not so busy during the daytime. It has not yet been observed whether they also feed at night in periods of bright moonlight.

This mode of life in semi-sessile colonies, unique among vertebrates, could presumably only have been developed because the eels find themselves in a sort of nutritive soup which is constantly flowing past them. Without any great demand for movement they are thus able to nourish themselves from the prey passing within their reach.

Giant plankton sieves

It strikes one as paradoxical that it should be the giants of the tropical oceans (and of the North Sea) which are plankton-eaters. Less selective in their manner of eating than the garden eels, which clearly take aim at their prey and then catch and eat it, the plankton-sieving creatures simply swim through the water with their mouths wide open and leave it to chance what they pick up. Once the water has had all the plankton screened out of it, it is discharged via the gill-slits. In the whale sharks the jaws are more than 2 metres (6½ feet) wide (Plate 132). Stewart Springer, a biologist of the US Fish and Wildlife Service, came upon a mass accumulation of crab larvae in the Gulf of Mexico during a voyage of the research ship *Oregon* one late afternoon in May 1953; this accumulation was being followed by several whale sharks. Each of them raised its huge head a couple of feet into the air, and then dived into the sea at intervals of 15–20 seconds, sucking in water in the process. Springer was able to observe clearly how small fish, even tunnies, were drawn into the current and swallowed.

I have encountered whale sharks many times in the Indian Ocean. As I swam near these gigantic bodies or held tightly on to the tail fins, bigger than a man – to give myself a wonderful tow through the ocean – I reflected on the enormous quantities of food that they must consume daily to satisfy their hunger. It has been calculated that the humpback whale (a mammal), living mostly on fish, would need some five thousand herrings for a single meal. Each herring would have had six or seven thousand copepods in its stomach. And if we now think how many algae each of the copepods must have consumed to eat its fill, we shall have an idea of the astronomical quantity of algae which have perished to make a meal for the whale.

In addition, the huge manta rays (Plate 128) and devil-fish sieve the plankton as they swim with widely-opened jaws, generally along the steep reef walls. They look like visitors from another era, and I have often wondered, looking at them, how these giants from a primitive age have managed to survive into our time.

But it is not only the giants of the fish world which sieve the pankton. In the Red Sea I have seen, generally towards evening, swarms of mackerel (*Sarda* sp.) only some 20 centimetres (8 inches) long patrolling up and down the reef edge in a tight group, with their jaws wide open, the rays of the setting sun reflected on their gleaming gill-covers. Once as I swam towards the shoal head-on, I could look right into the depths of those gaping throats. These fishes differ from their relatives – most of which are predators – in feeding on marine plankton, and by utilizing this source of nourishment they avoid competition with other species.

134 *(Opposite)* Schooling behaviour is a protective adaptation against the predatory enemy. It is impossible for the predator to fix his eye on an individual fish before snapping at it, because a succession of potential victims crosses his field of view, thus confusing his aim.

135

136

137

135–137 Off the reef swims a school of *Rhabdamia* (135). A predator arrives *(Aethaloperca rogaa)* and is hemmed in by the school of fish (136). The predator has no opportunity to attack and swims back to the reef (137).

138 Schooling fish (here, *Haemulon sciurus*) react to members of the same species by turning towards them. The members of the school recognize each other by a patterning specific to the species. A schooling reflex keeps them together.

138

140

142

139 Hunter and hunted. The large diadem sea-urchin *(Diadema setosum)* spends the daytime hidden or sitting in tight groups, with spines interlocked – this is a protective adaptation against their enemies. Only after nightfall, when their enemies are asleep, do the sea-urchins go foraging.

140–142 Some fish specialists feed on sea-urchins. Here a large triggerfish *(Balistes fuscus)* blows a powerful jet of water out of its mouth towards its victim (140). The sea-urchin loses balance and falls over, so exposing the vulnerable oral side. At this point the spines are shorter, so that the triggerfish can use its powerful teeth to bite open the urchin's body (141). In a very short time the viscera have been devoured (142). Nothing remains but the spines lying on the sea-bottom.

141

143–146 The small harlequin shrimp *(Hymenocera picta)* inhabits shady hiding-places in the shallow areas of the Indo-Pacific coral reefs. With its broadened flag-like antennae oscillating to and fro, it smells out its victims, preferably starfish. It follows them along the 'scent trail', then mounts the starfish, and with its claws supports its arm in a raised position, keeping its head down (143, 144). The victim is turned over onto its back (145, 146). Now the shrimp brings its small eating claws into play, and with them cuts deep wounds in its gradually-weakening victim. The preferred delicacy of the shrimp is the gonads. Aquarium experiments in Seewiesen have shown that this shrimp also attacks the destructive crown of thorns starfish. Whether it would be suitable for control of this creature on the reef remains to be proved (compare Plates 12–14).

144

146

145

148

◁147 A most unusual photograph: two large predators, a triachid shark or dogfish *(Triaenodon obesus)* and a large jack *(Caranx sansun)* hunting along the edge of the reef. They both work the same beat, preying upon fish which they take by surprise. The triachid generally lives near the bottom of the reef. When hunting, it swims along the steep reef wall. The fast-moving jack keeps up a continuous patrol along the edge of the reef and principally hunts fish swimming alone.

148 The large barracuda *(Sphyraena barracuda)* is a particularly feared predator. Barracudas, when young, swim in schools, but live in isolation as adults and move over to the reef. Attacks by this predator on human beings are triggered off by vigorous movements or by the presence of conspicuous, shining objects. In contrast to the shark, the barracuda only bites its victim once, tearing out huge lumps of flesh.

149 *(Overleaf)* The sharks of the high seas must keep continually moving to obtain a sufficiency of prey, since the region they inhabit is only thinly populated. Sharks primarily feed upon sick and weakly creatures and thus perform a function of natural culling among the species they prey upon. In this photograph an injured dolphin *(Tursiops truncatus)* is being attacked by a blue shark *(Prionace glauca)*.

150 *(Page 196)* Along the edge of deep coral reefs, as here where the continental shelf off Madagascar plunges into the depths, is the territory of the dangerous white-edge shark *(Carcharinus albimarginatus)*.

7 Schools and predators

During my observations of fishes along the reef it constantly strikes me that numerous species prefer to live as self-sufficient, unsocial loners, while others prefer swimming in pairs. Others again live in companionable groups, or in schools which give the impression of being a single huge organism.

What is it that holds these enormous swarms, consisting of thousands of creatures, together? As one brought up in the proper Darwinian tradition, one would also like to know what survival value this behaviour has and what adaptive value the formation of such swarms provides.

Observing a large school of defenceless sprats along the coral reef is quite a disturbing experience, even when we only look at them. The whole school seems to be constantly breaking up and reforming. If one fish for any reason starts off in a new direction, part of the crowd will follow it. The swarm of sprats is split up, and those left in the middle are for a moment in doubt as to where to go. Hesitantly they turn towards one half of the school, only to notice that the other half is in fact more attractive. In the meantime the leaders of the two groups have come together again. A minute or two later another fish at the other end of the school leaves the main track to pick up a titbit – a small snail which happens to be passing by – thus initiating another blind breakaway movement.

And so it goes on, day in, day out. A restless to-and-fro movement which often seems purposeless. And yet, I tell myself, nature must have arrived at the schooling drive by a process of selection. The first time I met this phenomenon I was prompted by the eternal indecision of the sprats to scatter the swarm violently. I played the part of an enemy and plunged savagely into the milling, silver-camouflaged mass of fish bodies. The whole group gave a single shudder, and the members of the school closed up and hemmed me in; had the fish been aggressive predators there would have been no escape. I was as it were in the middle of a planetarium, which changed its shape if I, the centre of this small universe, made the slightest movement. I had not the slightest chance of touching even the tail of a single fish.

With my diving gear on my back, my underwater note-pad before me, I am sitting at a depth of 16 metres (50 feet) on a block of coral, counting the number of attacks made by a damselfish (which I had marked) on other members of its species. There are eighty *Dascyllus trimaculatus* in a loose school above me, engaged in eating plankton. From time to time they shoot steeply upwards, because they have detected something eatable there. Suddenly the whole swarm of fish abruptly closes up and flees as a compact body to a coral stock. And here comes Sessie, a large, local barracuda, which I have known for some years. It is searching for prey.

A late afternoon along a steep reef wall: the rays of the sun slant through the water, illuminating the reddish body of the *Anthias* perch, thousands of which inhabit the edge of the reef. They maintain a station a couple of yards away from the reef and graze on the plankton. Frequently the males shoot out of the school and perform a conspicuous signal leap. The *Anthias* are very lively fish, which swim across the reef like a boiling cloud, in constant undulating movement. Abruptly, as though an order had been given, they disappear. Three large mackerel swim rapidly by.

At the lip of the continental shelf off the coast of Madagascar, along the edge of the precipice, is a level plateau of dying reef at a depth of 40 metres (130 feet). Thousands of *Lutianus* snappers swim in orderly formation across the reef. They are grazing on the growths of algae on the dying corals. All this presents a peaceful and harmonious picture, with no sign of hostilities between the creatures. Suddenly a shudder runs through the mass of fish and the snappers take up a tight formation near the sea-bed – undoubtedly an alarm reaction. I turn round and observe a large shark of the species *Carcharhinus albimarginatus*, which is slowly swimming up the precipice.

Here are three observations from different habitat zones in the coral reef. In the open water where there is no cover the sprats form a ball around their enemy; the *Anthias* along the edge of the reef find refuge in the fissures; a host of *Lutianus* snappers take up a tight formation on the sea-bed because the flat, dying coral reef provides insufficient cover.

The characteristic feature of all three examples is the communal reaction of the school to its enemy. Thus the school behaviour could in consequence be a protective adaptation to the enemy. But how then shall we explain the behaviour of the predators? Why do they not simply swim right into the school? After all, they only need to open their jaws and take their prey.

Associations between predator and prey

In the Red Sea I have observed for the past two years a home-ranging group of the cardinalfish *Rhabdamia* (Plates 135–137). This school lives at a depth of 14 metres (45 feet) above a small isolated coral reef. When I chase the fish, they swim with lightning rapidity into the reef as a body. Two brown reef perch *(Aethaloperca)* are their dreaded enemies – nevertheless they live in close proximity. If the perch are away from the reef for a short time, the cardinalfish will venture out a little further into the free water. However, as soon as the perch return, the cardinalfishes swim back to the reef and take up a position directed towards the rock face, ready for flight.

I have patiently sat for hours on this reef and watched the interplay between the hunter and the hunted. The reef perch do not have an opportunity to catch fish directly from the school. If they swim slowly out of the reef, the cardinalfish do not, as one might expect, take refuge in their hiding-holes; on the contrary, they make a circle around the enemy. They surround them in a milling scrum, exactly as I was hemmed in by the school of sprats. The predator's field of view is filled with a seething mass of fish bodies which flash before its eyes. But the predator must 'home in' on its victim before it attacks, and here there is no opportunity to do so – it is simply confused, and cannot make a bite.

The cardinalfish seem to be fully aware of this effect and give no indication of being in a hurry to run away from the reef perch. But the perch too have learned by experience that blindly swimming into the school is no better way of taking their prey. Nature has, however, shown them an expedient and taught them how they can nevertheless outwit the cardinalfish. From time to time they both unexpectedly rush at the swarm. The cardinalfish scatter, panic-stricken, and that short moment for which the school is no longer completely in control of the situation gives the predators their chance. The perch follow isolated, wildly fleeing fish which have become separated from the school. Well before they can rejoin the safety of the group, they have been caught and swallowed.

In many places in the Red Sea I have observed this unusual coexistence of predator and prey. The impression that I have gained is one not only of the protective function of schooling behaviour but also of the adaptation of the enemy to its prey. And an even more interesting thought struck me, namely that the school fish identified their enemy very well and behaved towards it in a different manner. From the reef perch the only danger they need expect arises from lightning darts of the predators. Against this the fish in the school protect themselves by boxing the enemy in tightly if an attack seems imminent.

Another enemy of the cardinalfishes is the lizardfish *Synodus,* a spindle-shaped bottom-dwelling predator. It is much smaller than the reef perch and can therefore slink close up to the school, without being immediately detected. Even before the school becomes uneasy and can begin to form its alarm circle around the enemy, the lizardfish is able to pick off some of the fishes around the edge of the group. *Synodus* is always a source of danger; consequently, as soon as the school discerns the enemy, they move a considerable distance away. The small predatory blenny of the genus *Runula* creeps up to the school and bites lumps out of the victim's skin. But once they see the attacker, the school encloses it in a hollow ball. On the other hand, the cleaner wrasse *Labroides dimidiatus* is a welcome visitor because it removes the skin parasites. This species is allowed to move about freely within the school.

It is a matter of intense interest when the individual predators each claim the area as their own and fight for the territory. The reef perches chase away the lizardfish which have found these school-inhabited rock cliffs. On one occasion I counted more than twenty-five individuals lying in ambush on the sea-bed. The reef perch does not even tolerate the small blenny *Runula* in its territory. In this way the density of predators is maintained at a constant level by the predators themselves – very much to the advantage of the cardinalfish. For them it is therefore much more advisable to remain in one place and pay their daily tribute to the reef perch, than to swim around freely and lose many more of their number to the attacks of numerous predators.

All for one, one for all

In a school of defenceless small fishes the constituent members are all equal. There is no leader, there are no officers and no non-conformists. Each respects the other, none is privileged above another. In addition, they have no reciprocal relationships; they form a large crowd, in which the individual remains anonymous.

From what has been said up till now, the reader might have gained the impression that schooling behaviour has been developed by nature solely as a protection against the enemies of the fish. This would, however, be a one-sided picture. Fish swarming does serve other functions. I have often seen mackerel, flutefishes, barracudas and even sharks hunting in groups. Eibl-Eibesfeldt has reported seeing a pack of grey sharks forcing a school of grey mullets into shallow water in a bay, in order to devour them. That is a piece of swarming behaviour aimed at obtaining food supplies. Schools can also be formed during the period of propagation of the species. This increases the probability of two members of the species meeting in a similar state of stimulation and of physiological readiness.

Nonetheless, the school appears to have arisen primarily under the selective pressure due to the 'enemy'. The prime means of achieving protection against this danger is the 'confusion effect'; this protection in fact relies in the last analysis on the incapacity of the predator to concentrate on a particular victim. Scientists have expressed the problem mathematically: the larger the school, the smaller the danger of being caught by a predator. The size of the swarm and the probability of being eaten by a predatory enemy are in approximately inverse proportion to one another. This is easily demonstrated from the example of a goldfish. If we put a few water-fleas into its bowl, it will eat them all, being able to concentrate on each in turn and catch them. One would assume that the goldfish would eat itself to death if faced with a host of waterfleas. But in fact the opposite occurs, and the number of creatures eaten remains constant, because the confusion effect is increased by the high density of the water-flea population.

Notwithstanding this, life in large aggregations does bring one disadvantage in its train. The larger the associative group, the more undecided are the members which compose it. This is something we have already learned in observing the swarm of sprats. Too many fish are able to swim off in different directions at the same time, so splitting up the school. However, their powerful social drive 'calls' them back to the swarm again, and it is only the short periods of indecision which constitute a danger.

Schooling behaviour not only offers advantages to the individual in the event of danger. Life in the school also makes it possible to achieve behavioural reactions which are not possible for the individual. Thus, for instance, goldfish learned more quickly to find their way through a maze if they had seen other members of the species swimming through the maze previously. Trained schools of carp were able to learn more quickly to evade a moving net than carp which had been individually trained. Learning is a special form of adaptation to surroundings.

A barracuda in flight

On one occasion I observed a large barracuda being attacked by a school of damselfishes because it had invaded their territory. Astonishingly enough the predator fled before the considerably smaller fishes. In the same colony there was also a large triggerfish which was chased away by a joint mock attack by the damselfish. This behaviour is called 'mobbing' by behavioural scientists. Konrad Lorenz cited the example of a hawk which had attacked a jackdaw, and was immediately mobbed by the flock of jackdaws to which the individual belonged. Although this did not serve to rescue the trapped jackdaw, this social defensive action can culminate in a situation where the hawk will be slightly less ready to hunt jackdaws than other birds – and contributes to the preservation of the jackdaw species. Even warning-off by an impressive but harmless mock attack can serve as a lesson to be more careful in future. Eibl-Eibesfeldt was the first to discover this behaviour in fishes. He observed schools of *Caesio* mobbing a moray eel and chasing it away. I have seen the same procedure with a large number of schooling fishes on many occasions. The most impressive was the 'mobbing' of a large octopus by damselfishes, when the octopus clearly intended to occupy the coral stock used by the damsels as a home. They continuously attacked the octopus, and initially it attempted to beat off the fish with its tentacles, but then speedily resorted to flight. These examples prove that creatures which are in themselves harmless can create an impression of danger when they are present in a mass.

What holds a school together?

If we observe the well-ordered movements of a school of fish, we shall soon wonder how it is that fish can react in so precise a manner, how they can transmit their moods and how they are able to behave as a composite organism. The position of the creatures within the school must be subject to continual correction. Most fish have a fixed distance between individuals, which is strictly maintained. Optical stimuli play a very important part in the orientation of an individual fish to its neighbours in the same school.

Several speculations have been put forward regarding the mechanisms which maintain the cohesion of the school, but no experimental proof has yet been provided. One thing is certain, however, namely that the fish possess a very strong schooling drive, a social instinct, and it is also known approximately in which part of the brain this is located. In a 1933 experiment, some minnows were operated on to remove the forebrain, and one operated minnow was returned to the group. The fish was able to eat normally and to swim normally, and its vision did not seem to be affected. But what turned out to be its fatal shortcoming was its behaviour towards its congeners. In fact, the surgical operation had removed every part of the brain responsible for triggering its social drive, among other things. The fish had lost its 'social relationship' to the members of the group. It no longer felt itself linked to them and swam heedlessly away – and the school blindly followed this 'brainless' leader. For species which live in groups the school is a living association, in which it feels itself protected. So it is not surprising that creatures accustomed to living in groups should become sick and generally die when isolated.

I have carried out underwater experiments to supplement these 'social relationship' tests. I built a dummy which I could draw through the water in any way I pleased. This dummy had to be as realistic as possible and possess all the visual characteristics of the species it represented, in order for the group to accept it. In actual fact I did succeed in taking schools of goatfish through the coral reef with me. They willingly followed their wooden comrade, which I drew behind me on a string. In this way I indirectly became the leader of the school of fish and held sway over innumerable goatfish, by imposing my will upon them through the medium of this dummy. A dangerous method of deliberately controlling depersonalized groups if one thinks of the significance of the 'social instinct'.

These experiments with operated fishes and with the dummy were carried out solely with individuals accustomed to social behaviour. They give an indication that fish are impelled by a social drive to react peaceably towards other members of their species. If this were the only mechanism underlying schooling behaviour, then the fish would always remain in a tight group. In other words, there must simultaneously be another force which serves to space the fish – just as electric charges of similar polarity repel each other. The regulation of the interval between neighbours in a school would then be a delicate balance between these two opposing forces. The functioning of each of these forces has been studied fairly recently by Evelyn Shaw, the American research worker. She isolated newly-hatched individuals of a gregarious species of fish, but enabled them from time to time to swim with fish accustomed to group living. The astonishing result of this experiment was that the isolated fishes did indeed draw towards each other and tried to swim on parallel courses, but that there was simultaneously a marked tendency for them to be repelled and swim away from each other. This behaviour was not exhibited by fish accustomed to swimming in schools. The isolated creatures constantly changed their position and performed a strange zig-zag dance – a piece of conflict behaviour, occasioned by the contrary drives – one prompting them to join the swarm, and the other to swim away from it.

Gregarious fish must therefore learn to achieve the right balance between these delicate mechanisms – which can only then guarantee the cohesion of the school, with all its advantages. Each fish must learn to swim in a group and needs social contact with its congeners for the purpose.

Predators of the sea

Hitherto I have been writing about the manner in which the predators and the prey are constantly being obliged to adopt new adaptations in the struggle for existence on the reef. The more discoveries the prey makes to use in its defence, the more difficult it will become for the predator to keep pace. In the course of the progressive conquest of the oceans, man is coming into increasingly frequent contact with the large marine predators, and must of necessity protect himself against them. He too, like a creature which is a potential prey, is obliged to adopt 'counter-adaptation'. Off the West African coast, professional divers have been attacked by huge groupers. Many cases have also been noted of large barracudas attacking human beings. Donald de Sylva analysed twenty-two cases, some of them having had fatal consequences, in which swimmers were attacked, generally at the surface of the water. In contrast to the shark, which takes several bites, the barracuda takes only one single bite at its prey, and this generally leads to its victim bleeding to death. However, the greatest potential danger threatening the diver comes from the shark.

The shark, 'the dangerous beast'

We have preconceived ideas about sharks and consider them to be repellent and 'wicked', purely because they possess certain characteristics which mark them out from the beginning as a 'beast': the slit-shaped, expressionless eyes, the head with the half-moon jaws and the terrifying teeth. From time immemorial man has drawn 'wicked' animals, bogies such as dragons, Medusa's heads and snakes with slitted eyes and jaws set with many teeth. The external appearance of the shark corresponds precisely to these preconceptions. Certainly one of the major reasons why we consider the shark to be an aggressive and bloodthirsty monster is that it is clearly our superior in the ocean. It is particularly among mariners that one finds this deeply-rooted hate against these predators.

Reports of cases where sharks have so mutilated human beings that they were no longer recognizable as such have nurtured this hate. A frightful example is the story of an American naval officer, Lieutenant-Commander Kabat, who in 1944 floated helplessly in the water off Guadalcanal for a whole night, being continually attacked by sharks: 'It came on again. I managed to hit it on the eyes and snout. The flesh of my left arm was torn into tatters . . . At intervals of ten to fifteen minutes it abandoned its slow progress and swam straight at me to attack me. Only twice did it swim underneath me. Since I was helpless against this type of attack, this was what I feared most. But as I lay so flat on the water, the shark could not get at me from underneath . . . The large toe of my left foot was hanging down by a thread. A piece was missing from my right heel. My elbow, left hand and left leg were torn. When it did not bite me, it rubbed great pieces of my skin off with its own rough skin. The salt water stemmed the flow of blood somewhat, and I did not feel any great pain.'

I shall now try to sketch a biological portrait of the shark, that will clear away the prejudices against it and will show us that its brutality towards its prey – which appears so appalling to us – has its justification.

Shark research

Every year some three hundred human beings fall prey to sharks. Probably the figure should be even higher, since not all these incidents are recorded. It is comprehensible, therefore, that we should intensify to an ever-increasing degree our research into shark behaviour. We have to develop reliable shark repellants, which will not only protect men working under water, but would be used, for example, in accidents involving international passenger ships and aircraft.

In April 1958, the American Institute of Biological Science set up in Washington an international centre for shark research, with the co-operation of thirty-four international experts; this is known as the Shark Research Panel, and its chairman is the well-known authority on sharks, Perry Gilbert. The purpose of this centre is, among other things, to study the biology and behaviour of sharks and to establish a statistical record and precise analysis of cases of attacks by sharks on human beings. Similar investigations are now being carried out in the Mote Marine Laboratory (previously Cape Haze Marine Laboratory) in Sarasota, Florida.

It is not only the behaviour of the shark towards human beings which is the subject of scientific research. Sharks are also coveted experimental material for medical research. Sharks resist cancer and do not suffer from heart diseases. Thus research workers at the National Institute of Health discovered in sharks certain types of serum antibodies, such as are formed in the human body in large quantities in cases of leukaemia.

The answer to the question of how the shark manufactures these antibodies could perhaps take cancer research one step further forward. For generations, preserved sharks have been much in demand as experimental material for demonstrations to students of anatomy. As a result we have a precise knowledge of the body structure of the shark, but know little about its behaviour. It is difficult to carry out research into the manner of life of the shark, because this can be done only in the open water, where the sharks themselves impose the experimental conditions.

Does this mean that every encounter with the king of the sea represents a grave danger?

Encounters with sharks

My most frequent meetings with sharks in the coral reef have occurred at the moment of first entering the water. They approach the diver inquisitively and then disappear equally suddenly. Any violent movement, and above all the noises of the breathing apparatus, will immediately put small coastal sharks to flight. As a result all I have seen of many sharks is the rapidly disappearing tail fins. Sharks in the open sea, on the other hand, are not so easily frightened by humans. They are sinister figures to us, which we contemplate with a mixture of fear and admiration as they swim rapidly past with powerful movements of their graceful bodies.

One rainy day I was diving off the west coast of Florida, together with Sylvia Earle, a research worker whose speciality is algae. In the cloudy water I could hardly see a couple of yards ahead and down near the sea-bed I rapidly lost sight of my companion. Suddenly I felt a strong pressure wave in the water. A lemon shark some 3 metres (10 feet) long was swimming towards me. Its movements were restless, and I knew from my experiences with this species at the Cape Haze Laboratory that I could expect to be attacked. The rapid appearance of the shark had scared me, and with a reflex movement I raised my underwater camera as though to protect myself. This set off the flash, and at that moment the half-moon shape of the shark's jaws and one piercing eye flashed by me in the turbid water. Had the flash put the shark to flight? Whatever the answer to that question might be, I promised myself never again to dive in turbid water.

One often hears the suggestion that sharks can be chased away by mock attacks. I was once diving with an Egyptian student, towards evening, at a depth of some 100 metres (330 feet) off a reef edge in the Red Sea. Suddenly, six medium-sized black-tip sharks – only 1½ metres (5 feet) long – came shooting up out of the deep water towards us, their tail fins threshing aggressively. In a flash they had surrounded us. The first made a forward movement, trying to jostle us. Waving our arms wide we swam towards it, to frighten it off. And indeed it did turn away shortly before reaching us. This was immediately followed by the attack from another shark, which we were also able to fend off. But the situation was becoming trickier by the minute, and the sharks' lunges were following at shorter and shorter intervals. As soon as the sharks retired a little, we rapidly swam towards the protection of the reef, only to lose a few yards in making another simulated attack. It was a cat-and-mouse game, which we did however eventually win. But we were far from feeling heroes once we reached the safety of the reef edge.

It is not always possible to summon up the courage needed to scare off a shark by a 'bluff' attack, above all not when the creature is longer than oneself. When I was diving once on a very cloudy day with a friend at a depth of 40 metres (130 feet) at the Banc d'Entrée in the Indian Ocean, a fully-grown hammerhead shark fully 6 metres (20 feet) long curved round ahead and suddenly swam towards us. The eyes on the hammer-shaped extensions moved in time with the slow to-and-fro motion of the head. Yard by yard it came toward us. Out there on the level ground there was no possibility of our finding cover. In the dark but clear water the shark was sinister and terrifying. I could not bring myself to swim towards it to deflect it from its course. In my fear I raised my film camera and pressed the release. It lasted eighteen seconds, the time for a complete film shot to run through, until the shark reached us, turned off shortly before contact and then slowly continued swimming along the reef wall. This encounter taught me that absolute quiet can be a very effective protection.

In my many encounters with sharks I have observed that most of them do not seem much impressed by human beings. However, I could never really decide when they were only annoyingly inquisitive and when they were really in an aggressive mood. Undoubtedly a shark which has first of all tested its potential victim by rubbing its rough skin against the prey will be ready subsequently to attack and eat it. But in some form or another this is the practice of almost all predatory animals, without our calling them particularly aggressive. What do we know nowadays of the life of the shark? Jacques Cousteau, who has met innumerable sharks in his many years of research diving, stresses that – however paradoxical this may seem – the more one has to do with the shark the less one knows about it.

The life of sharks

Some 250 species of salt- and freshwater sharks have been identified. In Lake Nicaragua in Central America there is the species *Carcharhinus nicaraguensis,* and in the Ganges, *C. gangeticus* – both species dangerous to man. Sharks have also been sighted repeatedly in the Zambezi. The shark family comprises a multiplicity of forms, ranging from the giants – the whale sharks – which are almost 20 metres (65 feet) long, to the dwarfs, measuring 15–16 centimetres (some 6 inches). Although they have a long evolutionary history, the sharks have successfully maintained their place to this day. In spite of their primitive jaw structure, they have never lost their predominant position as voracious predators of the sea and have even been able to resist the pressure of the more recent vertebrate fishes.

These predatory creatures have overcome the problem of competition for food by occupying different ecological niches in the ocean and hunting in different areas. In the chapter on the zones of the reef I have described how the sharks are found in various zones, so eliminating interspecific conflict.

The triachid shark is a bottom-dweller, while the grey and blacktip sharks keep to the open water off the reef. They try to avoid each other, because they occupy the same ecological niche. Along the edges of the reef precipices I have frequently observed the whitetip shark (*Carcharhinus albimarginatus,* Plate 150).

Some sharks exhibit astonishing adaptations to their habitat. For instance, the sharks of the open sea with their powerful, spindle-shaped bodies can swim fast and for long periods. This group contains the most aggressive and dangerous species such as the great white *(Carcharodon carcharias),* whitetip *(Carcharhinus longimanus)* and mako *(Isurus oxyrinchus)* sharks.

Since, as is well known, sharks do not have a swimbladder like other fish, they must continue swimming round tirelessly day and night to prevent themselves from sinking. These rapid-moving creatures are passive breathers, that is to say they employ the oxygen which enters the slightly opened jaw below the leading edge of the head and leaves by the gill-slit. This method of breathing is an excellent adaptation to life in the open water. However it only begins to function when the shark is swimming at a speed in excess of 2 metres (6½ feet) a second. Bottom-dwelling species must, therefore, employ active breathing movements.

The bottom-dwelling species are often coloured to match their background. The Wobbegong shark *(Orectolobus ogilbyi)* has even developed a certain patterning to provide it with good camouflage (Plate 72). The sand sharks, too, are difficult to see when they are on the bottom. The species *Carcharias taurus* has been responsible for the deaths of many swimmers, who failed to see this 5-metre (16-foot) shark even in shallow water. Many divers have attempted to awaken sleeping nurse sharks and have been surprised when they were attacked by these apparently harmless ground sharks. A case has been reported of a thirteen-year-old boy who was bitten in the arm by a nurse shark only 70 centimetres (27 inches) long because he had caught hold of its tail as it lay asleep.

Cousteau has made some very interesting observations on the behaviour of small reef sharks. By means of harpoons he attached plastic markers to the sharks' bodies, and was so able to show that they are attached to a particular spot. Coastal sharks even form group territories. Presumably there are hierarchies of rank within the groups – the weaker members avoiding the strongest (the alpha individual). The fact that they remain in one place contributes to making their hunting more successful.

Even for a shark it is not always an easy matter to catch its prey. Most reef fish have adapted themselves to the enemy which preys upon them and take refuge in their holes when it approaches. Consequently, sharks often hunt in packs and drive their prey. In the Red Sea I observed a territorial grey shark which had hit upon a quite remarkable idea. This shark hunted primarily along the edge of the reef below the breakers. At this point smallish fish are often caught in the undertow and are carried along by the waves. At this moment they are defenceless against a predator.

The easiest prey for sharks are the weakly and sick creatures; in this manner sharks carry out a natural culling of their prey (Plate 149).

Cannibals in the mother's body

Sharks either give birth to living young or lay eggs which have been fertilized within the body. The sharks' manner of copulation has been observed only rarely. Hans Hass has reported how male triachid sharks bit continuously into the gill-slits of a female in their attempts to grasp it. This behaviour was also noted by the U.S. ichthyologist Dr Eugenie Clark shortly before the copulation of a pair of lemon sharks. Viviparous sharks bear varying numbers of young; for example, the lemon shark has up to eleven, but the tiger shark can bear up to fifty. Can we take it that this difference in the number of young born is related to the manner of life of these creatures? Presumably the sharks of the open sea have to raise more young because many of them either starve in the infinite wastes of the open waters or are perhaps eaten by other sharks. Adult sharks frequently do not even spare their own young from their voracious appetites.

The sand shark *(Carcharias taurus)* has a most remarkable birth process. This species possesses two separate uteri, and only one youngster is born from each. On birth the young shark already has a most unusual form of struggle for survival behind it – in order to survive, it has had to eat up all the other young in the uterus. This instance of cannibalism in the mother's body must be quite unique in the animal kingdom. When Stewart Springer, who discovered this phenomenon, was examining a female sand shark, he was bitten in the hand by a young one.

Sharks and their prey

Investigations of the stomachs of sharks have repeatedly shown that these predators are not selective in the prey which they will consume. They eat virtually anything: fish, crabs, bivalves, cuttlefish, sea-birds, cans of food, and once a shark's stomach was found to contain a roll of roofing paper 9 metres long and 1 metre broad (30 feet by 3 feet). It is not surprising, in view of their predatory way of life, that every now and then human remains have been found in sharks' stomachs and have even led to the detection of crimes.

Marcus Coppleson, the Australian authority on sharks, is of the opinion that sharks which have once attacked swimmers in shallow water develop a special taste for human prey. A large white shark achieved a tragic notoriety off the coast of New Jersey in 1916, where within the space of ten days it had killed four human beings and severely wounded a boy. When it was caught two days later, its stomach still contained human flesh and bones.

Eugenie Clark has given reports of shark attacks off the west coast of Florida, which all occurred within a five-week period in the same region, although no human beings had been attacked by sharks in this coastal region for thirty-eight years. Probably these attacks too were the work of the same individual. These cases tend to support the hypothesis that sharks are easily led to concentrate on one particular prey.

A very widespread misconception is that sharks turn on their backs before biting. This is not true. The observations of Cousteau and others made in the open sea have indicated that sharks can eat in almost any position. Before they start biting, they will frequently swim several times around particularly large prey and will test it by bumping with the snout or rubbing it with their skin to see if it is tasty. Only then does there follow the attack proper, and this rises to a wild frenzy of feeding if several sharks are present together. Perry Gilbert writes that the shark brakes its impact with the prey by means of its pectoral fins – like an aeroplane coming in to land with the landing flaps down. Before making its bite, the shark raises its body slightly, opens its jaws wide (they can open wider than 90°) and buries its teeth in its prey with terrible force. Shaking its head to and fro several times, it will tear out pieces weighing from 5 to 10 kilograms (10 to 20 pounds). The pressure exerted in the bite was measured by Gilbert with a refined piece of apparatus, and was found to amount to several tons.

Blood, body juices and flapping movements of the prey increase the feeding stimuli of the sharks. Once they have begun a feeding frenzy, the creatures will stop at nothing and attack everything which comes within their range – including other sharks. On board the 'Rhincodon', the shark-catching boat of the Cape Haze Laboratory, we were often able to observe sharks in the water which had been caught on the hook and were being attacked by their congeners. A hooked shark in fact emits many of the characteristic stimuli of sharks' prey, and thus attracts the members of its own species. Often it happened that we only managed to land the heads of hammerhead sharks which we had hooked.

What attracts sharks?

Sharks are able to scent substances dissolved in water and carried by the currents at a distance of 350 metres (380 yards). This power of scent may at first sight appear phenomenal, but it is surpassed by the eel. One cubic centimetre of a given substance dissolved in fifty-eight times that volume of water at the sea bottom can be detected by eels in experiments.

Perry Gilbert presented lemon sharks with four externally identical cans pierced with holes, only one of which contained fish-meat. The test animals very rapidly identified this can. Tester, of Hawaii University, believes that sharks can even detect frightened or disturbed fish by means of a scent substance which these creatures are presumed to emit when in a state of fear. The experiments have shown that the sense of smell plays an important part in the shark's location of its prey.

Since sharks often hunt in turbid water or even in the unlit depths of the sea, it used to be thought that their power of sight was less powerful and was not employed in catching their prey. The eye of the shark does however exhibit certain anatomical peculiarities which enable it to make the maximum use of the smallest quantity of light. Behind the retina is a layer of reflecting guanine crystals, known as the tapetum lucidum. Once a ray of light has impinged on the light-sensitive points of the retina, it is reflected and thrown back by the tapetum, so that the retina is stimulated a second time by the same light rays. Many shallow-water sharks will for this reason protect their retina from excessively strong light, by dropping a kind of 'curtain' of pigmented cells over the tapetum.

Sharks are also attracted by flapping, irregular movements. When so attracted, they make a visual check of the places where they intend to bite. Cousteau's famous diver dummy Arthur was bitten in the leg during one attack. Bathers or men thrown out of wrecked ships, floating helplessly on the surface, emit – from a shark's point of view – ideal visual prey signals, and are consequently almost always bitten on arms and legs. The movements of swimming also simultaneously produce pressure waves which the sharks can detect with their lateral sensory organs. The perception of the pressure waves is of particular importance for the shark hunting in cloudy water and in the unlit depths of the open sea. Some time ago, scientists examined the sensitivity of the lateral sensory system. By surgical operation the faculties of hearing and sight were eliminated, but the sharks nevertheless reacted purposefully to certain phenomena in the water.

The hearing powers of sharks are also well developed. Only recently Dr Arthur A. Myrberg Jr., an American marine biologist, began experiments in the open sea off the Bahamas in which harmonic and non-harmonic tones of various frequencies were produced in the water. The reactions of the sharks were observed by means of an underwater television camera. The lower frequencies ranging from 20 to 100 cycles a second were the most effective and attracted the largest number of sharks. The upper threshold of perception lay around 800 cycles.

The Dutch research worker Sven Dijkgraf discovered that certain other sensory organs in the head, the so-called ampullae of Lorenzini can even detect electrical stimuli. To what end this faculty has been developed is not yet known. Can it also be a means for the sharks to locate living prey?

Defence against sharks

It was suggested some time ago that rotting shark meat and copper acetate had a repellent effect on dogfish. This led the US Navy to equip its men with copper acetate tablets for emergency use at sea. Hans Hass however subsequently demonstrated that most species of shark took very little notice of the copper acetate, since even harpooned fish which had been loaded with copper acetate were consumed.

During the Second World War, nigrosine dye was used as a shark repellant. Nigrosine spreads very rapidly in water and forms a large black cloud. This substance was thought to affect the sight of the sharks. And in actual fact the sharks did sheer away from the cloud of dye. They also did so when experiments were carried out with their noses blocked. This proved that the dye has no action on the sense of smell. Only when the sharks' eyes were covered with blinkers did they swim into the cloud.

Observations have shown that a shipwrecked mariner floating on the surface of the water is safer if his threshing, down-hanging legs cannot be seen by the shark, since they form a visual trigger stimulus which signals prey. A survival bag has been developed, into which the shipwrecked man climbs. It may also be that this method to some extent cushions the pressure waves produced by the swimming movements. The inventor, Scott Johnson, swam for a considerable time in such a survival sack among lemon sharks and was not attacked.

Hans Hass used to claim that sharks can be scared off by shouts. Many people diving for sport have used this method – some with success and some without. But it is quite certain that very few people have tried shouting at large sharks. Philippe Cousteau, son of Jacques Cousteau, considers the method to be ineffective; my experience, too, indicates that the larger sharks are not affected by shouts. Small coastal sharks, however, can be scared off either by shouting or by violent movements. Sharks are sensitive to movement. A stick with a nail in the end has been of great service to many divers to fend off shark attacks. Recently experiments have been carried out with electric repellent devices, which make use of the sensitivity of sharks to electric shocks.

The species of shark known with absolute certainty to be dangerous to human beings are the great white shark, mackerel shark, hammerhead shark, lemon shark, tiger shark and blue shark, the grey shark and the mako shark, and the oceanic whitetip shark. There are other species which are not reliably reported to have been responsible for attacks on human beings, but cannot be excluded from the list of possibly dangerous sharks – such as the blacktip shark of the Indo-Pacific. It is equally certain, however, that the majority of the 250 species of shark do not represent a danger to human beings.

The degree of aggressiveness of sharks depends on the time of day, the water temperature, the visibility and also on the manner of the swimmer's movement. Up to the present time, divers have been attacked less frequently than swimmers. This may be due to the fact that the diver moves much more quietly under water and is always prepared for an encounter with a shark. When a shark appears he therefore turns towards it. Swimmers at the surface, on the other hand, generally do not notice the approaching shark and do not exercise any control over their movements, so that they could more easily trigger off an attack.

During commercial and scientific investigations under the sea, protective cages into which the diver climbs have proved their worth. They do, however, restrict his radius of action considerably. It is my belief that if man recognizes the primacy of the shark in the ocean and accepts that on this occasion at any rate he is the inferior, the traditional hatred of sharks will slowly disappear. The frightening brutality of the manner in which sharks deal with their prey is to be understood as a behaviour pattern which has occurred as a result of selection.

The requisite protection against sharks for human beings cannot result in our exterminating sharks in blind rage, as was unsuccessfully attempted in Hawaii; on the contrary, behavioural studies must lead to the development of natural protective measures which can be of equal benefit to bathers and divers in tropical oceans and, no less, to the survivors of shipwrecks or aeroplane crashes.

The Photographers

Dr George J. Benjamin
Canada

Born 1918. Owner of the Benjamin Film Laboratories, Toronto.
Plate 40
taken with a Nikonos on CPS film.
Principal interests photography and speleology. Has investigated the Blue Holes in the Bahamas. His film *Andros Blue Holes* had its première in Santa Monica and New York (best documentary film of 1969). In 1970, collaborated with Jacques Cousteau in making a television film on the Blue Holes.

Jane Burton
England

Animal photographer.
Plate 24
taken with a Hasselblad 500C.
Has made numerous contributions to *Life, Reader's Digest, Animals.* Author of *Animals of East Africa* and *Life History of the Flamingos.* Lives with her family in Surrey (England), where she carries out studies in animal behaviour.

Ron Church
USA

Born 1934. President of SEACOR Inc., manufacturer of underwater photo equipment.
Plates 32, 75, 103
Awards: International Underwater Photographer of the Year 1963. Numerous national and international first prizes. Author of *Dans les Secrets de la Mer* (Secrets of the Seas); *Skin Diving in Hawaii; Surfing in Hawaii; Beginner's Guide to Photography Underwater;* publications in all leading American magazines. Has assisted Jacques Cousteau in his television series and books.

Benjamin Cropp
Australia

Born 1936. Film producer and photographer. Chairman of the Photographer Committees of the World Skindiving Federation and of the Australian Skindiving Federation.
Plate 128
taken with a Nikonos (28-mm.) on Kodak Ektachrome X film.
Underwater photographer of the year at Santa Monica, 1964. Winner of seven harpoon hunting competitions in Australia (prior to 1962); nowadays hunts solely with the camera. Contributed the photographs for the article 'Diving with sea snakes' in the *National Geographic Magazine* in April 1972. Author of *Shark Hunters* (New York, 1971). Producer of sixteen underwater films for television. His most recent film, *The Hungry Sea,* has also been shown in Europe.

Walter Deas
Australia

Born 1933. Underwater photographer.
Plates 25, 72, 114
taken with a Rolleimarin on Kodak Ektachrome film.
Australian underwater photographer of the year, 1969; gold medallist, British Film Festival, 1968; honorary member of Underwater Photographers' Hall of Fame, USA, 1969; silver medallist, Mondo Sommerso, Italy, 1971. Author of *Beneath Australian Seas; Seashells of Australia; Australian Fishes; Natural Life of the Barrier Reef; Coral Reefs of the Seychelles and the Great Barrier Reef.* Member of the Underwater Research Group of New South Wales and of the British Society of Underwater Photographers.

Professor Irenäus Eibl-Eibesfeldt
Germany

Born 1928. Professor of Zoology at University of Munich.
Plates 1, 4, 71, 86
taken with a Rolleimarin on Kodak Ektachrome Professional film.
Publications listed in the bibliography.

Hans Flaskamp
Germany

Born 1937. Architect.
Plates 52, 117
taken with Rollei and Hasselblad on Kodak Ektachrome film.

Dr Hans W. Fricke
Germany

Born 1941. Zoologist.
Plates 3, 5–9, 12, 14, 17–20, 30, 34–38, 42, 44–47, 62, 63, 65, 68, 70, 73, 76, 77, 79, 85, 88, 89, 91, 95, 98, 100, 102, 107–110, 115, 118, 122, 125–127, 136, 137, 139, 140, 142, 150
taken with Exa, Yashica, Hasselblad and Rolleiflex on Kodak Ektachrome Professional film.
Honoured by the *Bundestagung deutscher Sporttaucher.* Has been awarded prizes in the USA and in the German Democratic Republic. Publications listed in the bibliography.

Peter R. Gimbel
USA

Born 1928. Film producer and director, photographer, author.
Plate 149
Peter Gimbel's films are known throughout the world. His last full-length documentary film, *Blue Water, White Death,* was also shown with great success in Europe.

Hermann J. Gruhl
Germany

Born 1941. Economist.
Plates 15, 66, 105
taken with a Rolleiflex 3.5 F on Kodak Ektachrome X film.
Awards: Gold 'Camera Louis Boutan', Germany, several bronze 'Tritons', England and USA.
Has published many articles in international diving journals. Member of the *Verband deutscher Sporttaucher* and of the US Academy of Underwater Photographers.

Dr Hans-Rudolf Haefelfinger
Switzerland

Born 1929. Zoologist. High school teacher. Associate of the Naturhistorisches Museum of Basle and of the Laboratoire Arago, Banyuls-sur-Mer, France.
Plate 83
taken with a Nikonos on Ektachrome X film.
Awards: ORION, ISFA award for scientific films.
Publications: Vol I, 'The stinging coelenterates', of *Grzimeks Tierleben* (Grzimek's Animal Life). Articles in *Nautilus, Image, Palette,* etc.
Member of the scientific diving team of the Laboratoire Arago, Banyuls-sur-Mer, France.
Has made scientific films covering several fields of marine biology.

Dr Sebastian Holzberg
Germany

Born 1937. Biologist.
Plates 54, 58, 64, 69, 84, 87
taken with a Yashicamat using a Hecomar-I on Kodak Ektachrome Professional film.

Peter Kopp
Germany

Born 1934. Diving instructor.
Plates 27, 28, 67, 80, 82, 96, 133, 138, 148
taken with Rolleiflex and Hasselblad on Kodak Ektachrome film.
Awards: First Prize at 'Premio Sarra', Italy, 1969; absolute winner in all classes at 'Premio Sarra', Italy, 1970; 'Golden Trident' as best underwater photographer of the year, Italy, 1970; most successful underwater photographer in *Yearbook of Underwater Photography,* Santa Monica, 1970; special prize as best underwater photographer in all classes, Japanese Ministry of Tourism. Has published contributions in *Life, Paris Match, Epoca, Stern, Quick, Scala Int.* etc. Member of the Verband deutscher Sporttaucher.

Siegfried Köster
Germany

Born 1926. Chemical engineer.
Plates 33, 39, 50, 51, 57, 92, 111, 120, 121

taken with a Rolleimarin on Kodak Ektachrome X and Professional film. Author of *Welt unter Wasser* (Underwater World).

Professor Dietrich B. E. Magnus
Germany

Born 1916. Zoologist.
Plate 22
taken with Exakta Varex on Kodachrome film. Numerous publications in international specialist journals (see bibliography).

David Masry
Israel

Born 1943.
Marine biologist. Underwater photographer for the last five years.
Plates 16, 26
taken with a Nikon F on Kodachrome II film. Numerous prizes in national underwater photographic competitions. Has published photographs in international specialist journals.

Jack McKenney
Canada

Born 1938. Publisher of the periodical *Skin Diver.* Resident in USA.
Plate 56
taken with a Nikonos on Kodak Ektachrome X. Awards: four gold medals, silver and bronze awards; Midwest Underwater Photographer of the Year, 1971. Has published many articles and books. Member of the 'K-W Dolphins' and 'Martini's Outlaws' Clubs, Kansas, Missouri.

Horst Moosleitner
Austria

Born 1936. Teacher. Also a student of zoology and botany.
Plates 74, 81, 97, 99, 106
taken with a Proktisix on Kodak Ektachrome X film.
Awards: three gold medals, two silver and seven bronze medals in Rome, London, Los Angeles. Has published articles in diving and aquarists' periodicals. Author of *Leben unter Wasser* (Life Under Water).

Allan Power
Australia

Born 1933. Underwater photographer.
Plates 41, 90, 94, 104
taken with a Rolleiflex using a Rolleimarin on Kodak Ektachrome X film.
First prize 'Premio Levanto', Italy. Author of *Secrets of the Seas,* published by *Reader's Digest,* and has published in many international specialist journals.

Ruud Rozendaal
Netherlands

Born 1935. Aircraft design engineer.
Plates 29, 48, 55, 59
taken with a Rolleimarin on Kodak Ektachrome X and Professional film.
Awards: Premio Sarra, 1968 (bronze); 3rd International Festival of Underwater Photography, 1968 (bronze); 8th International Underwater Photographic Exhibition, USA, 1968 (gold and bronze); 4th International Festival of Underwater Photography, Brighton, 1970 (gold); British Sub-Aqua Club Trophy for the best colour transparency of the Festival. Author of *Fotojacht onder de Waterspiegel* (Hunting with a camera under water); *Alles over Sportduiken* (All about diving as a sport). Has contributed articles to *Delphin, Grzimeks Tierleben, Triton, Underwater Africa, Skin Diver, Yearbook of Underwater Photography,* etc.

Ludwig Sillner
Germany

Born 1923. Export sales executive.
Plates 10, 31, 93
taken with Rolleiflex 3.5 F using a Rolleimarin on Kodak Ektachrome Professional film.
Awards: Absolute winner, 'Premio Sarra', Rome, 1964; 'Golden Trident', Ustica, 1968; Photographer of the Year, Santa Monica, 1969, etc. Since 1960 has been a contributor to

five European diving journals. Author of *Kleiner Sprung ins große Meer* (A small leap into the vast ocean); *Mit der Kamera auf Unterwasserjagd* (Hunting underwater with a camera). Individual member of the Verband deutscher Sporttaucher. Underwater photographer on Commander Cousteau's ship *Calypso* in the Indian Ocean in 1967.

Piero Solaini
Italy

Born 1921. Technician.
Plate 119
taken with a Rolleiflex using a Rolleimarin on Kodak Ektachrome X film.
Awards: Absolute winner 'Maurizio Sarra', 1963; Premio Levanto; Premio città di Catania, etc.
Has had photographs published in specialist journals and in books.

Akira Tateishi
Japan

Born 1930. Underwater photographer.
Plates 112, 134
taken with Tateishi Bronica Marine and Bronica S2 on Kodak Ektachrome X and Fujicolor film.
Articles published in *Marine Park, Fishes, Marine Diving Magazine.* Member of the Japanese Photographic Society.

Valerie and Ron Taylor
Australia

Film producers, journalists.
Plates 2, 11, 13, 21, 32, 49, 60, 61, 101, 113, 123, 132
taken with Nikon F, Nikonos, Rolleiflex.
First prize at International Film Festival, Santa Monica, USA; first prize 'Maurizio Sarra', Italy.
Articles published in *National Geographic Magazine, Life, Reader's Digest, Mondo Sommerso, Skin Diver, Fathom.*

Herwarth Voigtmann
Germany

Born 1937. Sports instructor (skiing, diving).
Plates 53, 116, 124
taken using a Rolleimarin on Kodak Ektachrome Professional film.
Awards: 'Maurizio Sarra', 1967 (bronze).
Various articles in periodicals.

Bibliography

ABEL, E.: 'Fische zwischen Seeigel-Stacheln' ('Fish which live among sea-urchin spines') in *Natur und Volk,* 90, 2, pp. 33–8 (1960).

— 'Zur Kenntnis des Verhaltens und der Ökologie von Fischen an Korallenriffen bei Ghardaqa (Rotes Meer)' ('A contribution to the knowledge of the behaviour and ecology of coral reef fishes off Ghardaqa, Red Sea') in *Z. Morph. Ökol. Tiere,* 49, pp. 430–503 (1960).

BARDACH, J. E.: 'On the movements of certain Bermuda reef fishes' in *Ecology* 39 (1), pp. 139–46 (1958).

BOOLOOTIAN, R. A.: *The Physiology of Echinoderms.* New York and London, 1960.

COPPLESON, M.: *Shark Attack.* Sydney and London, 1958.

COTT, H. B.: *Adaptive Coloration in Animals.* London and New York, 1957.

COUSTEAU, J.-Y.: *The Silent World.* London and New York, 1953.

— *Life and Death in a Coral Sea.* London, 1971.

COUSTEAU, J.-Y. and PHILIPPE: *The Shark: splendid savage of the sea.* London and New York, 1970.

CROSSLAND, C.: 'The coral reefs at Ghardaqa, Red Sea' in *Proc. zool. Soc. Lond.,* Ser. A. 108, pp. 513–23 (1938).

Davenport, D., and K. Norris: 'Observations on the symbiosis of the sea anemone *Stoichactis* and the Pomacentrid fish *Amphiprion percula*' in *Biol. Bull.* 115, pp. 397–410 (1958).

Davis, W. P., and D. M. Cohen: 'A Gobiid fish and a palaemonid shrimp living on an antipatharian sea whip in the tropical Pacific' in *Bull. Mar. Science* 18 (4), pp. 749–61 (1968).

Eibl-Eibesfeldt, I.: 'Über Symbiosen, Parasitismus und andere zwischenartliche Beziehungen bei tropischen Meeresfischen' ('Symbiosis, parasitism and other interspecific relationships in tropical ocean fishes') in *Z. Tierpsychol.* 12, pp. 203–19 (1955).
— 'Der Fisch *Aspidontus taeniatus* als Nachahmer des Putzers *Labroides dimidiatus*' ('The fish *Aspidontus taeniatus* as a mimic of the cleaner wrasse *Labroides dimidiatus*') in *Z. Tierpsychol.* 16, pp. 19–25 (1959).
— 'Beobachtungen und Versuche an Anemonenfischen *(Amphiprion)* der Malediven und der Nikobaren' ('Observations and experiments on anemone fishes *(Amphiprion)* of the Maldive and Nicobar islands') in *Z. Tierpsychol* 17, pp. 1–10 (1960).
— 'Eine Symbiose zwischen Fischen *(Siphamia versicolor)* und Seeigeln' ('A symbiotic relationship between fishes *(Siphamia versicolor)* and sea-urchins') in *Z. Tierpsychol.* 18, pp. 56–9 (1961).
— 'Freiwasserbeobachtungen zur Deutung des Schwarmverhaltens verschiedener Fische' ('Open-sea observations as a means of explaining the schooling behaviour of various fishes') in *Z. Tierpsychol.* 19, pp. 165–82 (1962).
— *Land of a Thousand Atolls.* London, 1965.
— *Haie: Angriff, Abwehr, Arten* ('Sharks: attack behaviour, defence, species'), Stuttgart, 1965.
— *Grundriß der vergleichenden Verhaltensforschung* ('Basic principles of comparative behavioural research'). Munich, 1967.

Eibl-Eibesfeldt, I., and H. Hass: 'Erfahrungen mit Haien' ('Encounters with sharks') in *Z. Tierpsychol.* 16, pp. 739–46 (1959).

Eibl-Eibesfeldt, I., and G. Scheer: 'Das Brutpflegeverhalten eines weiblichen *Octopus aegina* GRAY' ('Parental care behaviour exhibited by a female *Octopus aegina* GRAY') in *Z. Tierpsychol.* 19, pp. 257–61 (1962).

Feder, H. M.: *Cleaning Symbiosis in the Marine Environment* ('Symbiosis', Vol. 1). New York, 1966.

Fishelson, L.: 'Observations on the biology and behaviour of Red Sea coral fishes' in *Bull. Sea Fish. Res. Stn. Haifa* 37, pp. 11–26 (1964).

Franzisket, L.: 'Die Stoffwechselintensität der Riffkorallen und ihre ökologische, phylogenetische und soziologische Bedeutung' ('Metabolic rate on the coral reefs and its ecological, phylogenetic and sociological significance') in *Z. vgl. Physiol.* 49, pp. 91–113 (1964).
— 'Riffkorallen können autotroph leben' ('Reef corals can live autotrophically') in *Naturwissenschaften* 56, p. 144 (1969).
— 'The atrophy of hermatypic reef corals maintained in darkness and their subsequent regeneration in light' in *Int. Revue ges. Hydrobiol.* 55, pp. 1–12 (1970).

Fricke, H. W.: 'Zum Verhalten des Putzerfisches *Labroides dimidiatus*' ('The behaviour of the cleaner wrasse *Labroides dimidiatus*') in *Z. Tierpsychol.* 23 (1), pp. 1–3 (1966).

— 'Der Nahrungserwerb des Gorgonenhauptes *Astroboa nuda*' ('Feeding pattern of the Basket stars *Astroboa nuda*') in *Natur und Museum* 96 (12), pp. 501–10 (1966).

— 'Garnelen als Kommensalen der tropischen Anemone *Discosoma*' ('Shrimps as commensals of the tropical anemone *Discosoma*') in *Natur und Museum* 97 (2), pp. 53–8 (1967).
— 'Beiträge zur Biologie der Gorgonenhäupter *Astroboa nuda* und *Astrophyton muricatum* (Ophiuroidea, Gorgoncephalidae)' ('Contributions to the zoology of the gorgon's-heads *Astroboa nuda* and *Astrophyton muricatum* (Ophiuroidea and Gorgoncephalidae)') in doctorate thesis to the Faculty of Mathematics and Natural Sciences at the Free University of Berlin, 1968.
— 'Zwischenartliche Beziehungen der tropischen Meerbarben *Pseudupeneus barberinus* und

Pseudupeneus macronema mit einigen anderen marinen Fischen' ('Interspecific relationships of the tropical goatfishes *Pseudupeneus barberinus* and *Pseudupeneus macronema* with certain other marine fishes') in *Natur und Museum* 100 (2), pp. 71–80 (1970).

FRICKE, H. W.: 'Neue kriechende Ctenophoren der Gattung *Coeloplana* aus Madagaskar' ('New creeping ctenophores of the genus *Coeloplana* from Madagascar') in *Mar. Biol.* 5 (3), pp. 225–38 (1970).

— 'Ein mimetisches Kollektiv – Beobachtungen an Fischschwärmen, die Seeigeln nachahmen' ('Collective mimicry – observations on fish schools which mimic sea-urchins') in *Mar. Biol.* 5 (4), pp. 307–14 (1970).

— 'Die ökologische Spezialisierung der Eidechse *Cryptoblepharus boutoni cognatus* für das Leben in der Gezeitenzone' ('The ecological specialization of the lizard *Cryptoblepharus boutoni cognatus* for life in the tidal zone') in *Oecologica* 5, pp. 380–91 (1970).

— 'Lebensweise des im Sand lebenden Schlangensterne *Amphioplus* sp.' ('On the behaviour and pattern of life of the sand-dwelling brittle-star *Amphioplus* sp.') in *Helgoländer wiss. Meeresunters.* 21, pp. 124–33 (1970).

— 'Ökologische und verhaltensbiologische Beobachtungen an den Röhrenaalen *Gorgasia sillneri* und *Taenioconger hassi*' ('Ecological and behaviour-biological observations on the garden eels *Gorgasia sillneri* and *Taenioconger hassi*') in *Z. Tierpsychol.* 27, pp. 1076–99 (1970).

— 'Fische als Feinde tropischer Seeigel' ('Fishes as enemies of tropical sea-urchins') in *Mar. Biol.* 9 (4), pp. 328–38 (1971).

FRICKE, H. W., and M. HENTSCHEL: 'Die Garnelen-Seeigel-Partnerschaft – eine Untersuchung der optischen Orientierung der Garnele' ('The partnership between shrimp and sea-urchin – a study of the optical orientation of shrimps') in *Z. Tierpsychol.* 28, pp. 453–62 (1971).

GERLACH, S. A.: 'Über das Korallenriff als Lebensraum' ('The tropical coral reef as a habitat') in *Verh. Dt. Zool. Ges. Münster (Westf.)*, pp. 356–63 (1959).

GILBERT, P. W.: 'The behaviour of sharks' in *Sci. Amer.* 207 (1962).

GOHAR, H. A. F.: 'Commensalism between fish and anemone with a description of the eggs of *Amphiprion bicinctus* RUPPEL; in *Publ. Mar. Biol. Stat. Ghardaqa, Egypt,* 6, pp. 35–44 (1948).

GOREAU, T. F.: 'The ecology of a Jamaican coral reef. I. Species composition and zonation' in *Ecology* 40, pp. 67–90 (1959).

— 'Growth and calcium carbonate deposition in reef corals' in *Endeavour* 20, p. 32 (1961).

— 'Calcium carbonate deposition by coralline algae and corals in relation to their roles as reef-builders' in *Anm. N.Y. Acad. Sci.* 109, pp. 127–67 (1963).

GRAEFE, G.: 'Die Anemonen-Fisch-Symbiose und ihre Grundlage nach Freilanduntersuchungen bei Eilat, Rotes Meer' ('Anemone-fish symbiosis and its basis, studied in the wild off Eilat, Red Sea') in *Naturwiss.* 50, p. 410 (1963).

HASS, H.: *Men and Sharks.* London, 1954.

— *Under the Red Sea.* London, 1953.

— *We Come from the Sea.* London, 1958.

— *Expedition into the Unknown.* London, 1965.

— *In unberührte Tiefen* ('Into unexplored depths'). Vienna, 1971.

HEARLD, E. S.: *Living Fishes of the World.* New York and London, 1961.

HIATT, R. W., and D. W. STRASBURG: 'Ecological relationships of the fish fauna on coral reefs of the Marshall Islands' in *Ecol. Monogr.* 30, pp. 65–127 (1960).

KAESTNER, A.: *Lehrbuch der speziellen Zoologie. I. Wirbellose* ('Handbook of specialized zoology. I. Invertebrates'). Stuttgart, 1963.

KLAUSEWITZ, W.: 'Biologische Bedeutung der Färbung der Korallenfische' ('The biological significance of the coloration of coral fishes') in *Zool. Anz. Suppl.* 22, pp. 329–33 (1959).

— 'Das Farbkleid der Korallenfische' ('The coloured livery of the coral fishes') in *Natur und Volk* 91 (6), pp. 204–15 (1962).

— 'Über einige Bewegungsweisen der Schlammspringer *Periophthalmus*' ('Certain patterns of movement in the mud-skipper *Periophthalmus*') in *Natur und Museum* 97 (6), pp. 211–22 (1967).

KLAUSEWITZ, W., and I. EIBL-EIBESFELDT: 'Neue Röhrenaale von den Malediven und Nikobaren (Pisces, Apodes, Heterocongridae)' ('New Garden eels from the Maldive and Nicobar Islands (Pisces; Apodes; Heterocongridae)') in *Senck. Biol.* 40, pp. 135–53 (1959).

KRAMER, E.: 'Zur Form und Funktion des Lokomotionsapparates der Fische' ('The form and function of the locomotor apparatus of fishes') in *Z. wiss. Zool.* 163, pp. 1–2 (1960).

LIMBAUGH, C.: 'Cleaning symbiosis' in *Sci. Amer.* 205, pp. 42–9 (1961).

LIMBAUGH, C., M. PEDERSEN and F. A. CHACE: 'Shrimps that clean fishes' in *Bull. Marine Sci. Gulf Caribbean* II, pp. 237–57 (1961).

LINSENMAIR, K. E.: 'Konstruktion und Signalfunktion der Sandpyramide der Reiterkrabbe *Ocypode saratan*' ('The construction and function of the sand pyramids built by the ghost crab *Ocypode saratan*') in *Forsk. Z. Tierpsychol.* 24, pp. 403–56 (1967).

LONGLEY, W. H. and S. F. HILDEBRAND: 'Systematic catalogue of the fishes of Tortugas, Florida' in *Papers Tortugas Lab.* 34, pp. 1–331 (1941).

LORENZ, K.: 'Naturschönheit und Daseinskampf' ('The beauty of nature and the struggle for existence') in *Kosmos* 58, pp. 340–8 (1962).
— 'The function of colour in coral reef fishes' in *Proc. Roy. Inst. Great Brit.* 39, pp. 282–96 (1962).
— *On Aggression.* London and New York, 1966.
— 'Über die Entstehung von Mannigfaltigkeit' ('On the origin of multiple forms') in *Die Naturwissenschaften* 12, pp. 319–29.

LUTHER, W.: 'Symbiose von Fischen (Gobiidae) mit einem Krebs *(Alpheus djiboutensis)* im Roten Meer' ('Symbiosis between fishes (Gobiidae) and a shrimp *(Alpheus djiboutensis)* in the Red Sea') in *Z. Tierpsychol.* 15, pp. 175–7 (1958).

MAGNUS, D. B. E.: 'Wasserströmung und Nahrungserwerb bei Stachelhäutern des Roten Meers' ('Water current flow and feeding in echinoderms of the Red Sea') in *Ber. Phy. Ned. Ges. Würzburg* 71, pp. 128–41 (1962–4).
— 'Zum Problem der Partnerschaften mit Diadema-Seeigeln' ('The problem of partnership with the sea-urchin *Diadema*') in *Verh. dtsch. zool. Ges. Munich* 1963, *Zool. Anz. Suppl.* 27, pp. 404–17 (1964).
— 'Zur Ökologie einer nachtaktiven Flachwasserseefeder *(Octocorallia pennatularia)* im Roten Meer' ('On the ecology of a nocturnal shallow-water sea-pen *(Octocorallia pennatularia)* in the Red Sea') in *Veröff. Instit. f. Meeresforsch. Bremerhaven* 369, p. 380 (1966).
— 'Bewegungsweisen des amphibischen Schleimfisches *Lophalticus kirkii magnusi* KLAUSEWITZ (Pisces, Salariidae)' ('Movement patterns of the amphibian blenny *Lophalticus kirkii magnusi* KLAUSEWITZ (Pisces, Salariidae)') in *Biotop. Verh. dtsch. zool. Ges. Jena 1965. Zool. Anz. Suppl.* 29, pp. 542–55 (1966).
— 'Zur Ökologie sedimentbewohnender Alpheus-Garnelen des Roten Meeres' ('On the ecology of sediment-dwelling Alpheus shrimps of the Red Sea') in *Helgoländer wiss. Meeresunters.* 15, pp. 506–22 (1967).

MAGNUS, D. B. E., and U. HAACKER: 'Zum Phänomen der ortsunstäten Ruheversammlungen der Strandschnecke *Planaxis sulcatus* BORN' ('The phenomenon of the variable-situation resting groupings of the beach snail *Planaxis sulcatus* BORN') in *Sarsia* 34, pp. 137–48 (1968).

MCGINITIE, G. E. and N.: *Natural History of Marine Animals.* London and New York, 1949.

ODUM, H. T. and E. P.: 'Trophic structure and productivity of a windward community on Eniwetok Atoll' in *Ecol. Monogr.* 25, pp. 291–320 (1955).

PHILLIPS, C.: *The Captive Sea.* London, 1964.

RANDALL, J. E.: 'Fishes of the Gilbert Islands' in *Atoll Res. Bull.* 47, pp. 1–243 (1955).
— 'Fish service stations' in *Sea Frontiers* 8, pp. 40–7 (1962).
— 'Food habits of reef fishes of the West Indies' in *Stud. Trop. Oceanogr.* 5, pp. 665–847 (1967).

RANDALL, J. E., and W. D. HARTMANN: 'Sponge-feeding fishes of the West Indies' in *Mar. Biol.* 1, pp. 216–25 (1968).

RANDALL, J. E. and H. A.: 'Examples of mimicry and protective resemblance in tropical marine fishes' in *Bull. Marine Sci. Gulf Caribbean* 10, pp. 444–80 (1960).

RANDALL, J. E. and H. A., R. E. SCHROEDER and W. A. STARCK: 'Notes on the biology of the echinoid *Diadema antillarum*' in *Carib. J. Sci.* 4 (2, 3), pp. 421–33 (1964).

RICKETTS, E. F., and J. CALVIN: *Between Pacific Tides.* Stanford, 1962.

RIEDL, R.: *Fauna und Flora der Adria.* Berlin, 1963.

ROUGHLEY, T. C.: *Wonders of the Great Barrier Reef.* Sydney and London, 1936.

SCHEER, G.: 'Der Lebensraum der Riffkorallen' ('The habitats of the coral reef') in *Ber. 1959/60 Nat. wiss. Ver. Darmstadt,* pp. 29–44 (1960).
— 'Viviparie bei Steinkorallen' ('Viviparity in stony-corals') in *Naturwiss.* 47, 10, pp. 238–9 (1960).

SCHLICHTER, D.: 'Das Zusammenleben von Riffanemonen und Anemonenfischen' ('Symbiosis of reef anemones and anemone-fishes') in *Z. Tierpsychol.* 25, pp. 933–54 (1968).
— 'Chemischer Nachweis der Übernahme anemoneneigener Schutzstoffe durch Anemonenfische' ('Chemical demonstration of appropriation of specific substances from anemones by anemonefishes') in *Naturwissenschaften* 57 (6), pp. 312–13 (1970).

SETCHELL, W.: 'Biotic cementation in coral reefs' in *Proc. nat. Acad. Sci. Washington* 16, pp. 781–3 (1930).

SHAW, E.: *Schooling in Fishes: critique and review in development and evolution of behaviour.* San Francisco, 1970.

STRASBURG, D. W.: 'Notes on the diet and correlating structures of some central Pacific echeneid fishes' in *Copeia* 3, pp. 244–8 (1959).

SYLVA, D. P. DE: 'Systematics and life history of the great barracuda' in *Stud. Trop. Oceanogr.* 1, pp. 1–153 (1963).

TINBERGEN, N.: *A Study of Instinct.* Oxford, 1961.
— *Social Behaviour in Animals.* London and New York, 1953.

VAUGHAN, T. W.: 'Corals and the formation of coral reefs' in *Amer. Rep. Smithsonian Inst. 1917,* pp. 189–276 (1919).

VERWEY, J.: 'Coral reef studies. 1. The symbiosis between Damselfishes and Sea-Anemones in Batavia Bay' in *Treubia* 12, pp. 305–66 (1930).

WAHLERT, G. V.: 'Die ökologische und evolutorische Bedeutung der Fischen-schwärme' ('The ecological and evolutionary significance of fish schooling') in *Veröff. Inst. Meeresforschung Bremerhaven* 8, pp. 197–213 (1963).

WAHLERT, G. V. and H. V.: 'Beobachtungen an Fischschwärmen' ('Observations on fish schools') in *Veröff. Inst. Meeresforschung Bremerhaven* 8, pp. 151–62 (1963).

WEBER, E.: 'Über Ruhelagen von Fischen' ('The resting positions of Fishes') in *Z. Tierpsychol.* 18, pp. 517–33 (1961).

WICKLER, W.: 'Zum Problem der Signalbildung am Beispiel der Verhaltens-Mimikry zwischen *Aspidontus* und *Labroides* (Pisces, Acanthopterygii)' ('The problem of signalling, using the behaviour mimicry of *Aspidontus* and *Labroides* as an example') in *Z. Tierpsychol.* 20, pp. 657–79 (1963).
— 'Natürliche Augen- und Eiattrappen an Fischen: innerartliche Mimikry als Sonder-funktion der Körperfärbung' ('Natural eye and egg dummies in fishes: intraspecific mimicry as special function of body coloration') in *Veröff. Inst. Meeresforschung Bremerhaven,* third special volume, pp. 222–7 (1963).

WICKLER, W., and U. SEIBT: 'Das Verhalten von *Hymenocera picta* DANA, einer seesterne-fressenden Garnele ('The behaviour of *Hymenocera picta* DANA, a starfish-eating shrimp') in *Z. Tierpsychol.* 27, pp. 352–68 (1970).

WIENS, H. J.: *Atoll Environment and Ecology.* New Haven and London, 1962.

WINN, H. E.: *The Biological Significance of Fish Sounds: Marine Bio-Acoustics.* Oxford, 1964.

YONGE, C. M.: *A Year on the Great Barrier Reef.* London and New York, 1930.
— 'Studies on the physiology of corals. V. The effect of starvation on the relationship between corals and zooxanthellae' in *Gt. Barrier Reef Exped. 1928–29, Sci. Rep.* 1, pp. 117–211 (1931).

ZUMPE, D.: 'Laboratory observations on the aggressive behaviour of some butterfly fishes' in *Z. Tierpsychol.* 22 (2), pp. 226–36 (1965).

Index

Numbers in italics refer to plates